Advanced Principles of Effective e-Learning

Books in this series

Principles of Effective Online Teaching
Nicole A. Buzzetto-More, Ed.

Advanced Principles of Effective e-Learning
Nicole A. Buzzetto-More, Ed.

Advanced Principles of Effective e-Learning

Edited by

Nicole A. Buzzetto-More

Informing Science Press

Advanced Principles of Effective e-Learning

ISBN-10: 1-932886-02-8
ISBN-13: 978-1932-88602-3

Published by

Informing Science Press
publishing arm of the Informing Science Institute:

131 Brookhill Court
Santa Rosa
California
95409
USA
Phone: +1 707 531 4925
Fax: +1 480 247 5724
ISPress.org
InformingScience.org

Printed in the US, Canada, the Netherlands, United Kingdom, and Australia

This book is dedicated to

Alan and Catherine Buzzetto

Acknowledgements

This book has been made possible by the creative and dedicated efforts, not only of the authors, the international board of reviewers whose suggestions and editorial comments were invaluable, Eli Cohen from Informing Science Press, and Retta Sweat-Guy whose editorial contributions were priceless.

Advanced Principles of Effective e-Learning

Contents

Buzzetto-More, N. A. (Ed.). (2007). *Advanced Principles of Effective e-Learning*
Santa Rosa, California: Informing Science Press.

Preface

Advance Principles of Effective e-Learning

For educators and trainers alike e-learning has become a familiar part of our vernacular that has forever changed our constructs of the 'class-room' as networked multimedia have been able to extend and redefine the teaching and learning experience. As we have stumbled through the development and integration of e-learning into academia we have been required to face and overcome numerous obstacles. These experiences have yielded valuable insights, as well as useful, solutions. This text has brought together the knowledge and experience of a host of experts who will share a variety of best practices for the development, management, sustenance, and assessment of e-learning endeavors. It also includes practical suggestions for evaluating assessing the quality of e-learning, as well as using e-learning for program and student assessment. This book is designed for administrators, key decision makers, educators experienced with e-learning, and instructional technology students. It marries the leading literature and prevailing ideologies with practical solutions and real-world examples.

Hundreds of proposals were submitted and considered and all drafts were double blind reviewed. This book represents the top ten percent of all submissions.

Dr. Nicole A. Buzzetto-More
University of Maryland Eastern Shore
December 2006

Buzzetto-More, N. A. (Ed.). (2007). *Advanced Principles of Effective e-Learning*
Santa Rosa, California: Informing Science Press. (pp. 1-25)

Chapter 1

The Six C's Framework for E-learning

M.R. (Ruth) De Villiers

Introduction

The 'technology' within *digital learning technologies* should be subservient to the 'learning'. Technology is a tool and a medium, but not the message itself. A step towards this is the development of e-learning based on sound *learning theory* (De Villiers 2003, 2005). Reeves & Reeves (1997) take a stance as they present various *pedagogical dimensions* by which Web-based instructional environments can be characterized. It is important that explicit pedagogy be integrated into learning resources and that content and methods be directly related to the learning objectives and outcomes (Clark & Mayer, 2003; Duchastel, 1998; Firdyiwek, 1999; Winn, 1999). In fact, Winn (1997) advocates that theoretical components should be included in instructional technology curricula. He points out that approaches which are purely prescriptive, i.e. 'how-to' models, are inappropriate, because human behavior is not predictable. Different learners may be taught by the same instructional method, yet their performances may vary considerably. Learning strategies vary as well. Learners with metacognitive ability are able to monitor their learning process and change strategies when they feel that an alternative approach is more effective.

With the belief that learning theory should contribute to the delivery and practice of learning, this chapter presents a framework that can assist educators and designers in developing e-learning resources that support flexibility, yet are based on solid theoretical foundations. The Hexa-C Metamodel (HCMm) is a synthesis of contemporary learning theories and practices, whose six elements: cognitive learning, constructivism, components, creativity, customization, and collaborative learning, should play a role in the design and development of e-learning.

The Six C's can also be used in evaluating educational applications and web-based learning from the perspective of learning theory.

Background

As a background, we briefly consider certain concepts relevant to the relationship between learning theory and actual learning or instruction. These factors are the shift in pedagogical paradigms, the role of technology, and various classifications of content domains.

Parallel with the evolution of educational technology, is the shift in pedagogical paradigms and learning theories through *behaviorism*, *cognitivism* and *constructivism*, two of which are more thoroughly addressed in later sections of this chapter. Behavioral instructional design models have roots in the programmed learning of the 1960s and are fairly rigid (Dick & Carey, 1996). Behaviorism is related to objectivism as it sets out to achieve defined objectives. It places importance on measurable, observable, performance-based outcomes, and uses stimuli to evoke the desired response in the learner, followed by reinforcement. Behaviorally-oriented instruction presents information in a systematic, individualized, but program-controlled, manner.

The cognitive approach emphasises the mental processes involved in learning, whereby learners acquire various information structures and procedures. The exploratory cognitive view treats learning less systematically, as the system presents phenomena that learners investigate by interaction and transformations, e.g. Logo (Papert, 1980).

At the end of the spectrum is the constructivist ethos emanating from pioneers such as Inhelder & Piaget (1958) and Bruner (1967; 1999). This involves an open-ended, flexible, exploratory view, which situates learning contextually, encouraging learner-centricity and active generation of products by learners themselves. Learning is scaffolded, rather than tutored, and the learner's role is highly participative.

Another approach to instruction and learning is the pragmatic view that has no explicit theoretical foundation and is done in a practical needs-driven way. Other perspectives are *hybrids*, including features from various theories and models, in line with suggestions by Hannafin, Hannafin, Land and Oliver (1997) that strategies need not necessarily originate from the same theoretical base.

Forms of e-learning vary according to the use and role of technology as a tutor, an environment or a tool. The role being played by technology in a particular situation affects the choice of an applicable learning-theory foundation. Zucchermaglia (1991, cited by Winn 1992) distinguishes between full and empty technologies. A full technology teaches defined content. It contains information to be transferred interactively to learners; for example, a computer-based tutorial, as described by Alessi & Trollip (2001). An empty technology, by contrast, is not an instructional system, but a tool or shell that can accept content. It supports exploration and construction, e.g. searches and generation of products using the Internet, as well as use by learners of commercial software packages for manipulation, documentation, presentation and communication.

The kind of content to be learned is another major consideration in determining appropriate learning-theory foundations. There is a difference between well-structured domains and ill-structured domains (De Villiers, 2003; 2005; Hannafin, Land & Oliver, 1999; Jonassen, 1999; Landa, 1998). The content in well-structured (or closed) domains consists of tightly defined concepts, e.g. syntactic, scientific and computational material, where rules, procedures or algorithms are prescribed. Ill-structured (or open) domains, on the other hand, are characterized by problems with multiple solutions and various alternatives, e.g. the social sciences, management sciences, environmental disciplines and design disciplines, which require reflective practice (Schön, 1987), expert-type knowledge, and heuristic processes.

In another classification of content, Simon (1981) distinguishes between the *natural sciences* and *design sciences*. Natural sciences relate to natural phenomena such as the sciences of physics, astronomy, biology, etc. Their associated descriptive theories set out the laws and relationships governing these systems. Design sciences, on the other hand (also called sciences of the artificial), relate to man-made applied disciplines, such as medicine, engineering, architecture and instruction, for which associated prescriptive theories and models set out procedures for accomplishing ends. Apt as this classification may be in distinguishing between the two kinds of disciplines, designers must exercise discernment when using so-called prescriptive theories, since they may not be context-sensitive and may work against design flexibility.

The Hexa-C Metamodel

With this background, it is clear that the choice of a theoretical foundation for an e-learning resource depends on various factors. No single paradigm is right for all situations – content domain, subject matter, target learners, type of assessment and the physical or virtual environment all play roles. Each application has its own focus. Furthermore, technology should be matched to the pedagogical philosophy, as patterns of teaching and learning are identified (Shneiderman, Borkowski, Alavi & Norman, 1998).

The Hexa-C Metamodel (HCMm) (De Villiers, 1999; 2003; 2005; De Villiers & Cronjé, 2005) is an approach that integrates six selected inter-related concepts from contemporary learning theory into a framework that serves both as a design aid and as an evaluation approach for investigating existing resources from the perspective of learning theory. It is called 'Hexa-C', because each of its six elements starts with the letter 'C', and 'Metamodel', because it is a synthesis of existing theories, models and paradigms, including certain classic writings. It is not the idea that every e-learning resource or system should conform to all six HCMm elements. When developing or investigating a particular environment or artifact, the educator, instructional designer, or evaluator should carefully consider the six elements, determining which are relevant to the content and context. They should then consider to what extent, and in what ways, those elements could be optimally implemented.

Three of the Cs: *constructivism*, *cognitive learning* and *components*, are mainly theoretical aspects, while the others: *collaborative learning*, *creativity*, and *customization*, are practical methods used by educators to foster effective and affective learning. Figure 1 shows the hexagonal framework of the HCMm, representing its inter-related elements as merging segments around the hub of technology as a means of delivery and emphasizing the importance of environment by embedding the whole within context.

Following Figure 1, the main features of the six C-elements are outlined and their implications are explained. For those who wish to delve more deeply, references are provided to original authors.

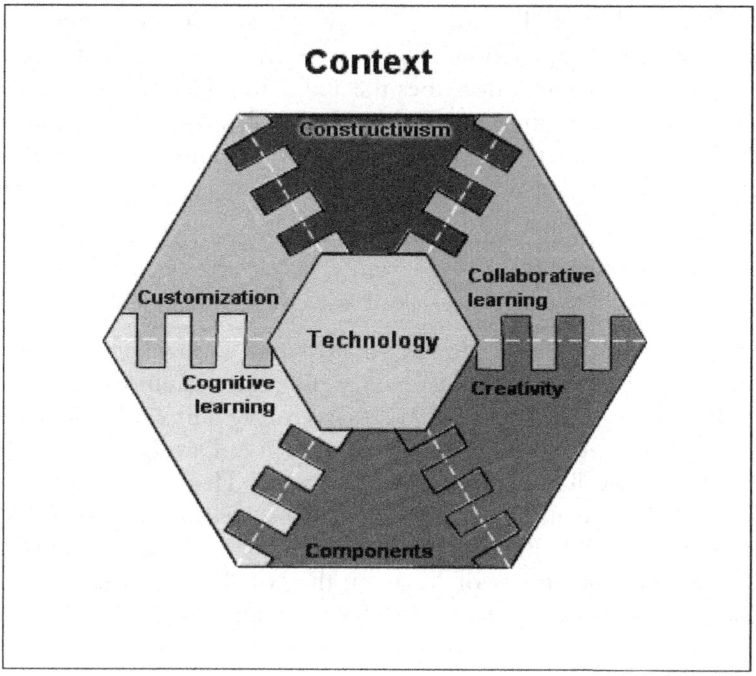

Figure 1. The framework of the Hexa-C Metamodel

1. Cognitive learning

Learning should support cognition and the formation of internal knowledge structures that represent causality and other relationships between concepts. If this occurs effectively, it helps retention and also facilitates transfer as we learn skills in other domains. Cognitivism relates to the process of learning, in particular, to cognitive processes such as analytical reasoning, the formation of internal schemata and mental models, and critical-and-creative thinking. In the 1980s and 1990s researchers such as Newell and Simon (1972) and Gagné and Glaser (1987) put forward various human information processing models to represent human comprehension, learning, the acquisition of knowledge, and problem-solving.

The cognitive approach views the process of learning and the cultivation of cognition as more important than the generation of deliverables and products by learners. New knowledge encountered should be integrated with prior learning, so that we build new skills on prior learning. Educators in the cognitive school encourage learners to develop critical thinking skills and to reflect on their learning. Metacognition – which means cognition about cognition or, more simply, thinking about thinking – is a term that describes the ability to understand and evaluate one's own cognitive processes. Learners who acquire these kinds of skills find it easier to approach the planning, monitoring, and self-regulation of learning.

Educators use various strategies to support cognitive learning. Authentic problem solving plays a role and another approach is the explicit teaching of cognitive strategies alongside content knowledge.

Some of the researchers who have contributed to the development and growth of the cognitive paradigm, including classic authors, are: Anderson (1983), Gagné & Merrill (1990), Greeno, Collins & Resnick (1996), Inhelder & Piaget (1958), Jonassen, Campbell & Davidson (1994), Minsky (1975), Newell & Simon (1972), Osman & Hannafin (1992), Reigeluth (1999), Reigeluth & Moore (1999), West, Farmer & Wolff (1991), and Winn (1990). Bloom's (1956) *Taxonomy of Educational Objectives* identifies six major kinds of learning, the last three of which, namely: anlysis, synthesis, and evaluation, call for cognitive skills.

2. Constructivism

Constructivism relates to personal knowledge construction and interpretation, active learning, and anchored instruction. The presentation of multiple perspectives on an issue provides a broad base for insight and interpretation, as does the use of different media and modes. Constructivist mechanisms include problem/project-based learning, open-ended learning environments, flexible learning within ill-structured domains, and authentic tasks – without simplification of complexity.

Constructivism is non-prescriptive; it is not direct instruction, but rather aims to set up learner-centric environments and activities. Instead of using educational technology to teach content and emphasize performance, constructivism uses e-learning to promote participative learning, active involvement and situated learning that can be applied in the real world. Closely associated are collaborative activities and research by learners, using a variety of resources (Bruner, 1967; 1999; Duffy & Jonassen, 1991; Hannafin et al, 1997; Jonassen 1994; 1999; Mayer, 1999; Reeves & Reeves, 1997; Winn, 1992).

Constructivism is a controversial approach, which led to various debates in the early 1990s. One of the contentious aspects is assessment.

Constructivists such as Cunningham (1992) and Jonassen (1992) propose integrated assessment – goal-free needs assessment embedded in multi-facetted and complex contexts or the evaluation of skills used in problem solving. Evaluation should be flexible and less criterion-referenced, and testing in decontextualized settings should be largely avoided. Goals of learning can be negotiated with learners and the resulting objectives can be used in guiding learning and assessment. Evaluation can include self-evaluation.

There are also diverging opinions on the role that errors play. It is a constructivist tenet that people learn from their mistakes (Squires & Preece, 1999). Constructivism views errors as largely transitional and beneficial; therefore it doubts the value of approaches that help learners avoid mistakes. Mistakes can help learners establish tentative beliefs, which are challenged and may change when the learner encounters new information (Hannafin, 1992). According to Lebow (1993), errors can be viewed as positive stimulants that create disequilibrium, leading to reflection and restructuring. Finally, constructivist approaches can lead to 'cognitive conflict' and 'cognitive complexity' (Perkins, 1991) when educators withhold explicit teaching and direct solutions. Some learners (and some educators) find this disturbing and prefer systematic teaching.

While the early constructivist-behaviorist debates strongly promoted principles of constructivist learning, they did not always propose explicit and practical ways of implementing constructivism. The very idea of 'constructivist instructional design models' was a contradiction in terms! By the mid to late 1990s, however, there was an impact on practice with the advent of constructivist design and practical suggestions for the use of technology for constructivist learning (Cennamo, Abell & Chung, 1996; Hannafin, Land & Oliver, 1999; Jonassen, 1999; Jonassen & Rohrer-Murphy, 1999; Land & Greene, 2000; Lebow, 1993; Mayer, 1999; Willis, 2000; Willis & Wright, 2000).

3. Components of learning

Components relate to the basic knowledge, skills and methods of a domain. There are unitary components and composite components, which combine the units. While it is clear that the HCMm has a cognitive bias, its framework includes components of learning and instruction (Reigeluth, 1999) in the belief that some types of skills and learning are best communicated by an instructivist approach that decontextualizes certain knowledge in order to teach effectively.

One approach is component display theory (CDT) (Merrill, 1983), which suggests that there are different kinds of learning outcomes and that different kinds of internal and external conditions are necessary to promote each kind of knowledge- or skill acquisition. CDT is based on relationships between the kind of content taught (*fact*, *concept*, *procedure*, and *principle*) and the level of performance required (*remember*, *use*, or *find*). Merrill defines the four types of content: 'facts' are arbitrary pieces of unitary information, while 'concepts' are groups of objects, events or symbols sharing a common characteristic. A 'procedure' is an ordered set of steps to accomplish a goal or solve a particular problem, and 'principles' are correlational relationships that interpret or predict a situation. With regard to performances, 'remember' requires a learner to recognize, then to reproduce, items of information. 'Use' involves application of some abstraction or approach, and 'find' requires the derivation or synthesis of a new abstraction or process. Despite its cognitive and constructivist bias, the HCMm framework includes components of learning and instruction in the firm belief that some types of skills and learning are best taught in this manner, sometimes even in a decontextualized setting.

CDT examines whether the instructional strategies used in a learning event can effectively achieve its instructional goals. Each learning objective is related to the appropriate content and desired performance, resulting in an instructional component that is positioned on a two-dimensional performance-content grid. Merrill (2001) emphasizes the use of instructional components as theoretical tools in the design of effective, efficient, and appealing instructional products, both in directive tutorial material and in environments for experiential learning.

One might wonder – does instruction in the basic components conflict with personal knowledge construction and interpretation? No, in many situations, the two approaches are complementary and compatible. Particularly in well-structured domains with tightly defined concepts, initial explicit teaching of the basic knowledge, concepts, and notations is fundamental. Such foundational knowledge and skills provide a platform for subsequent interpretation and synthesis.

4. Creativity and motivation

Designers should aim to incorporate affective aspects within learning, seeking creative and innovative instructional strategies, aiming for novelty within functionality in ways that motivate learners intrinsically (Dick 1995; Jones, 1998; Malone, 1981). The motivational ARCS Model of Keller & Suzuki (1988) advocates gaining attention, ensuring relevance, instilling confidence, and achieving learner-satisfaction. While intrinsic motivation is essential to gain and maintain learner engagement, there are situations where extrinsic motivation is appropriate.

Certain cognitive psychology research into the cognitive and affective domains, shows that the cognitive and emotional subsystems of the brain are highly interrelated (Price, 1998; Wager, 1998). As a result, there is increasing awareness of the importance of motivating learners, and educators should capitalize on this cognitive-affective connection.

Caropreso and Couch (1996) stress the value of creativity and innovation in workplace training. Although there are designers of instructional and learning products who are naturally creative, Caropreso and Couch believe that the necessary creative skills can be learned. A successful creative product should engage learners, ideally inducing flow (Csikszentmihalyi, 1990). 'Flow' is an optimal experience where participants concentrate deeply and thoroughly enjoy the activity. Learners experiencing flow tend to forget that they are actually involved in a learning experience! Most important, creativity in instruction engenders creativity in learners (De Villiers, 2005), as novel learning situations stimulate learners to lateral thinking (de Bono, 1977).

There is a notable convergence between the disciplines of human-computer interaction (HCI) on the one hand, and instructional design and learning theory on the other. If the aspect of usability is important for professionals in the workplace, how much more essential it is to attain high usability and supportive interaction in e-learning, where learners are first exposed to a web site or courseware before they can even begin to engage with the learning activities. To support and motivate learners, HCI principles regarding interface design, interaction design, and usability (Dix, Finlay, Abowd & Beale, 2004; Preece, Rogers & Sharpe, 2002; Shneiderman, 1998) should be studied and applied.

5. Customization

The movement towards customizing, or individualizing, learning (Alessi & Trollip, 2001; Bruner, 1967; Norman & Spohrer, 1996; Reigeluth, 1999) aims for learner-centric instruction that supports individual learners. The idea is that learning resources should adapt to individual profiles, match learners' needs and interests, support personal learning processes, and foster the synthesis of original products by learners. Learners should be able to take initiative regarding (some of) the methods, time, place, content, and sequence of learning.

The methods and extent of customization vary according to the instructional paradigm supported. In behaviorist products and in some of the cognitive educational software, customization refers to adaptable systems with branched program control over the material presented and its sequence. In cognitive situations that tend towards constructivism, there is increased learner control. Learners may be able to manage their own progress and may select learning activities. In true constructivism, learning is even more open, and customization occurs as goals are negotiated, with learners possibly even selecting the topic/s they investigate, as well as the approach and resources.

An interesting form of customization occurs when learners use an artifact or tool in an original manner – unintended by the designer – often with a positive effect. This is termed incorporated subversion (Squires, 1997).

6. Collaboration

Collaborative learning involves joint work, social negotiation, a team approach, accountability, and peer evaluation, i.e. sharing of responsibility within a group. It optimizes complementary abilities and instills collaborative skills in learners (Nelson, 1999; Panitz, 1996; Singhanayok & Hooper, 1998). Its leading proponents in the 1990s, Johnson & Johnson (1991), describe the key elements of cooperative and collaborative learning as being shared goals, positive interdependence, individual and joint accountability, the ability to communicate, interpersonal skills, empowerment of learners and cooperative assessment.

Rules and allocation of roles can capitalize on strengths and sidestep weaknesses. Teamwork demands effective communication skills between participants, and is excellent preparation for the workplace.

Duffy and Jonassen (1991) point out that collaborative learning often occurs within constructivism. Nelson (1999) discusses collaborative problem solving, describing how groups of learners tackle tasks requiring complex knowledge and skills. This expertise can be combined in different ways to complete the tasks successfully.

Before the days of the Internet, collaboration was mainly restricted to class-based teaching and other face-to-face situations. This has been transformed as the World Wide Web and digital technologies change the face of learning and communication, bringing spatial independence and offering a variety of forms of synchronous and asynchronous collaboration.

Implementation of the Six Cs

Learning is complex. It depends on the content to be taught and learned; it is also context-specific, being influenced by the environment and culture in which it is situated. What works in one situation may not transfer directly to another. This section re-visits each of the six C's of the Hexa-C Metamodel and suggests practical ways of applying it and implementing it in e-learning, with due regard to the type of domain and learning content. The suggestions involve generic principles and discussion and do not include highly specific examples. It is up to the reader to contextualize the concepts to his/her own situation. Each of these application subsections should be studied along with its theoretical partner in the preceding section.

First, we briefly review well- and ill-structured domains, as introduced in the earlier 'Background' section and touch on appropriate forms of technology. Well-structured, or closed, domains have well-defined rules and procedures. Some suitable e-learning applications are computer-based tutorials, practice environments, custom-designed hyperlinked 'textbooks', and simulations. In general, these are full interactive technologies, also outlined in the earlier section. In order to be classified as e-learning and not just e-teaching, they must involve more than passive transmission of information. They should be highly interactive and dynamic in the sense that usage and system response/feedback differs from learner to learner. Ill-structured domains lend themselves to more flexible e-learning, often with emptier technologies, focusing on exploration, communication and construction. Typical examples are open learning environments (OLEs) (Hannafin, Hall, Land & Hill, 1994; Hannafin et al, 1999) and constructivist learning environments

(CLEs) (Hannafin et al, 1997; Jonassen, 1999), as well as project-based and problem-based learning (PBL) (Savery & Duffy, 1995). Furthermore, learners should find material via World Wide Web searches or by accessing databases/intranets, and should use computing tools independently to manipulate information and generate their own products, which can be uploaded to web portfolios. Some of the particular requirements of these two contrasting types of domains are addressed further in the following subsections on practical ways of applying the Six C's.

Cognitive learning: Application

Cognitivism, which straddles the divide between objectivism and constructivism, stresses the integration of new knowledge with prior learning. Knowledge in structural and procedural domains can be presented incrementally, with background knowledge accessible on demand. A step-wise, chunked approach can legitimately be adopted – possibly alternating teaching, examples and illustrations with questions and participative learner-activities. The relationship between theory and practice may be addressed deductively – introducing concepts and rules before applying them, or inductively – moving from examples towards identification of underlying theory and concepts. Problem solving in closed domains mainly uses the information provided, rather than material sourced independently by the learner. Learners' answers to questions should ideally receive individualized responses in the form of diagnostic feedback. Learners should be stimulated to provide input to parameter-driven animations or dynamic simulations which, in turn, should provide rapid visual reactions to foster cognition. Advanced 'challenge' activities or complex exercises should be provided to stimulate higher-order thinking skills (HOTS).

When technology is applied in ill-structured, open domains, learners work in a less tutored manner as they investigate issues with various interpretations and approaches. Foundational knowledge should be accessible as scaffolding, while new knowledge should be available on a need-to-know basis. Such connections can be implemented by hyperlinking. To strengthen the cognitive-affective connection, the attention and relevance aspects of the ARCS Model are important. Where practical, schedules should support self-paced learning, accompanied by planning and self-monitoring. Learners should make independent use of HOTS, and should apply creative and critical thinking as they undertake authentic activities and develop deliverables. Immediate online diagnostic assessment is unlikely to occur, but supportive feedback (online or offline) should be provided by facilitators.

Constructivism: Application

In closed domains, personal knowledge construction is the antithesis of componential instruction, yet the two, as stated in an earlier subsection, are compatible and complementary. Constructivist approaches can be used for exploration and discovery learning beyond the basics. E-learning environments can facilitate this, for example, interactive problem solving can be facilitated by scaffolding showing generic steps in the problem-solving or decision-making process. Multiple presentation modes (audio, video, graphic, textual, interactive, etc.) and varying perspectives on the content, support learners in gaining insight. Providing non-examples – as well as examples – and requiring learner-synthesis of examples are two means of contextualizing understanding. In the social-constructivist approach, collaborative activities can help to consolidate concepts.

Constructivist tasks are characterized by authenticity, often simulating a real-world problem and providing information on a just-in-time basis. As stated, these approaches and the lack of explicit teaching can lead to cognitive conflict and complexity, termed constructivist frustration by the present author. For some learners this may be inadvisable, in fact, it may not work. But for those who persevere, there is retention and transfer.

Constructivist assessment is complex. It is not criterion-referenced but aims to assess learning gain, incorporating multimodality, portfolios, journal-keeping, self-evaluation, and peer-evaluation. In some cases, there may be no 'threatening' grading mechanisms at all.

The resource-rich environment of the World Wide Web is ideal for the implementation of constructivism in open domains with alternative correct approaches. There is a natural relationship between constructivism and the freedom of the Web (Hedberg, Brown & Arrighi, 1997). Personal goals and learning objectives can be negotiated. In OLEs, CLEs and PBL – all mentioned in the introductory part of this section on application of the C's – the problem (or question) frequently drives the learning and learners generate personal interpretations and solutions. Web Quests are guided Web activities, which are more structured than open-ended searches. The educator sets up a series of links to specific sites on the WWW in order to address a topic or elicit answers to questions. Web Quests were developed by Dodge (1996) to foster data gathering, interpretation, deliberating, and the drawing of conclusions.

Exploration of resources results in discovery learning. The presentation of authentic contextualized problems helps to avoid the pitfall of simplifying complexity, and leads to experiential learning as information is accessed on a need-to-know basis. Exploration of erroneous paths leads to reconfiguration, reflection and restructuring. A word of caution is expressed regarding electronic searches. Resources should be scrutinized for authenticity, validity, and reliability. They should be assessed against Smith's (1997) criteria of accuracy, authority, currency, uniqueness, external links, and quality of the writing. Post graduate learners should be advised to use accredited and reviewed sources only.

In another form of e-learning, software tools can be used by learners to manipulate and present information. For example, they can use spreadsheets as cognitive tools to display findings and manipulate multiple parameters. In design disciplines, graphics packages and animation can be used to convey information. Learners can develop multi-media products and create websites on which to post personal learning products and portfolios.

Real-world activities enforce standards beyond the norm for academic efforts, thus demanding superior efforts. These can lead to the constructivist learner-frustration mentioned earlier. But in some cases, research and development projects conducted by learners, particularly by postgraduate students, become real-world products beyond academia, usable in the workplace and in the market.

There are some drawbacks, however. True constructivism entails freedom from constraints – a lack of boundaries, which can impact negatively on other aspects. For example, self-paced work can obstruct collaboration when one team member delays the others' schedules. This is particularly difficult in distance collaboration, where offenders can disappear into cyberspace, beyond chasing and catching!

Components of Learning: Application

Though outcomes-based education may be flavor-of-the-day, procedural domains usually require explicit teaching of basic knowledge and skills, which all learners should grasp. Tutoring in unitary components, such as the facts and concepts of component display theory (CDT), should precede the more composite components, such as the CDT principles and procedures. Furthermore, demonstrations of examples involving algorithmic knowledge can be subdivided into steps.

Examples and exercises can themselves be available as components and selected by learners, with learner-control over the quantity and degree of difficulty. In well-structured domains, interactive practice environments can be designed to offer far more than the original computer-based, program-controlled drill. Visual templates outlining generic steps within fixed procedures can support learners while they are doing exercises. Where appropriate, basic components, such as relevant theory and definitions, can be linked to the task on hand. In less-structured domains, the relevant components should be a pre-existing foundation, known by learners before they undertake open-ended problem solving. There can be linked electronic access to basic resources or to lists of external resources. Furthermore, components can be contextualized and integrated as required, unlike conventional teaching of basic knowledge- and skill components, where facts and concepts might be introduced in a decontextualized manner.

Creativity and Motivation: Application

Creativity in e-learning applications for closed domains is closely related to intrinsic motivation and engagement of learners. Innovative analogies, metaphors, and icons can serve as bridges to the familiar and enhance e-learning applications that provide many resources in a single environment. They can incorporate informality and fun. The attributes of attention, relevance, confidence, and satisfaction, proposed by the classic ARCS Model, are as appropriate to web-based learning as they were to the educational technology of the 1980s and 1990s. Multimedia presentations are means of novelty and engagement, which can also offer multi-sensory reinforcement. Be careful of using special effects, 'edutainment', 'bells and whistles' that could distract. Technological novelty should be used in ways that serve the learning experience and do not detract from it.

The truly creative designer will pay attention to the human needs of learners as well as to the application of technology. Creativity is really rewarded when learners experience 'flow', forgetting time and tackling more than envisaged! Intrinsic motivation relates to aspects of the learning experience that naturally motivate and engage learners, but the achievement-oriented learner may also require extrinsic motivation and recognition of performance, over and above the instruction or learning. Another human factor to consider is that, as part of the affective-cognitive connection, values and emotions influence learners' initial approach to accessing learning and acquiring knowledge (external affective aspects) as well as their continued attitude and attention (internal affective aspects). Computer-based and web-based learning resources are computing applications and also course material. E-learning designers are therefore both software developers and educators. As such, they should know how to apply usability principles and guidelines from the discipline of human-computer interaction (HCI).

The foregoing is equally applicable to open domains, but there are further corollaries. First, we must also consider the technological activities/projects/products required from learners and make sure that there is adequate technical training and support. Failure to do this may demotivate them. With adequate skills on the part of the learner, open learning systems and web 'classrooms' have scope for vibrant human interaction, informality and excellent deliverables – both solo products and team efforts. Research conducted using the HCMm framework (De Villiers & Cronjé, 2005) shows that creativity in the learning experience fosters creativity among learners. Novel situations stimulate them to rise to the occasion.

Second, important means of motivation are content and values. Real-world problem solving and authentic tasks may lead to content being the motivator rather than artificial means. When adults undertake continuing education or market-related training, educators should use the context of the learners' workplaces where possible. They should be required to develop workplace-related artifacts and do assignments that lead to career development. By exposing mature learners to professional-type situations, they take ownership of the problem. Values serve as motivators where activities are vision-driven and relate to the learner's personal passions (De Villiers & Queiros, 2003).

Customization: Application

Classic CAI incorporated branching, a form of program control that navigates learners through material according to their performance in a pretest. Learner-controlled customization, by contrast, allows learners some say regarding content, components, time, and place of learning. In structured domains a defined learning sequence may be required, so learner control over content should be used circumspectly. Multiple activities and presentation means may be optimized as learners choose between tasks, modes, and media, learning asynchronously in the time, pace and (sometimes) the place of their choice. When material is available to be used in a learner-controlled manner, it allows learners to use time in a way that supports their personal learning styles and to re-view and re-do as required. Varied levels of difficulty and degrees of help should be available – offering challenge or support as required for learners' styles or stage of learning. Flexible use patterns can lead to some surprises as learners apply originality and incorporated subversion, customizing an artifact or environment in an unintended manner for their own purposes.

Learner-control and learner-centricity come into their own in unstructured domains where learners tackle open-ended projects. Auto-customization and customization-by-content occur as learners either:

- Choose their own approach within set content, using tools and techniques of choice;
- Determine own content and direction within broad domains, developing personal products; or
- Customize learning by taking specialist roles in collaborative teams.

In problem-based contexts, learners can personalize the experience by directing their learning; conducting independent research; taking specialized roles in teams; collecting, analyzing and manipulating information; and drawing their own conclusions. Findings can be presented – online or face-to-face – using verbal communication, printed format, graphic displays, video, slide shows, or a combination.

Collaboration: Application

Collaborative learning is not usually considered suitable for algorithmic tasks with tightly defined procedures, yet spontaneous joint use occurs of e-learning applications intended for individual use in such domains. Co-operative problem solving, two-at-a-computer, is an effective means of learning and building confidence. Peer-teaching of technology occurs naturally, helping to alleviate the problems of learners who have gaps in their technological expertise.

For electronic collaboration in domains with open content, we consider two cases. First, e-learning within contact-learning provides an ideal infrastructure for teamwork, joint projects, and collaborative problem-solving. Role allocation should capitalize on skills and strengths. Teamwork is an effective and efficient method of handling heuristic tasks requiring complex knowledge and varied expertise; it helps to prepare learners for the real world and the workplace. The second case, electronic distance collaboration, has greater complexities, but communication can be implemented using e-mail, bulletin boards, chat facilities, online conferencing and discussion forums. Distance teamwork lacks the factors of urgency and human dynamics inherent in face-to-face contact (De Villiers & Cronjé, 2005). Ways and means must be sought to compensate for this, such as clear definition of roles and rules. In both face-to-face and distance-learning, collaborative assignments and products can be posted on web sites in a way that acknowledges the role played by each member of the team.

Challenge Exercises: from Theory to Application to Real-World Practice

Consider the content you teach or plan to teach. Maybe it is introductory computer programming at a university; perhaps it is 20th century literature for high school students, viewed against the background of history of the time, or maybe it is something quite different – like teaching first graders to read. Classify the content according to its domain. Discuss its context and the target learners. Then propose two e-learning applications that would support your target group in learning the required content and skills in ways that include higher-order skills and that encourage critical and creative thinking, where applicable. Explain which HCMm elements the e-learning applications would implement and how.

Now take another, quite different, domain and content, perhaps something you studied yourself, and hated! Place yourself in the role of a teacher or facilitator for this course and do the exercise above.

Conclusion

The Hexa-C Metamodel has been presented as a conceptual framework that integrates tenets of contemporary learning theory. Synergy occurs between its elements, the six C's, as evidenced by the explicit interrelationships mentioned in several sections. The design, development, implementation and evaluation of e-learning can be enhanced by considering the HCMm elements and applying those appropriate to the context. No single paradigm is appropriate – no 'one size fits all' – but its elements can be translated into principles, design guidelines, and evaluation criteria for different domains and subject matter. The instructional designer, e-learning practitioner or educational web developer should ensure that technology serves as a hub that delivers the message and does not distract or detract from the message.

References

Alessi, S.M. & Trollip, S.R. (2001). *Multimedia for learning: methods and development.* Boston, MA: Allyn and Bacon.

Anderson, J.R. (1983). *The architecture of cognition.* Cambridge, MA: Harvard University Press.

Bloom, B.S. (1956). *Taxonomy of educational objectives. Handbook 1: Cognitive domain.* London: Longmans.

Bruner, J.S. (1967). *Toward a theory of instruction.* Cambridge, MA: Harvard University Press.

Bruner, J. (1999). *Constructivist theory.* Retrieved September 2006, from http://tip.psychology.org/bruner.html

Caropreso, E.J. & Couch, R.A. (1996). Creativity and innovation in instructional design and development: the individual in the workplace. *Educational Technology, 36*(6):31-39.

Cennamo, K.S., Abell, S.K. & Chung, M. (1996). A "Layers of Negotiation" Model for Designing Constructivist Learning Materials. *Educational Technology, 36*(4):39-48.

Clark, R.C. & Mayer, R.E. (2003). *E-learning and the science of instruction.* San Francisco, CA: Pfeiffer.

Csikszentmihalyi, M. (1990). *Flow: the psychology of optimal experience.* New York: Harper and Row.

Cunningham, D.J. (1992). Assessing constructions and constructing assessments: a dialogue. In: Duffy, T.M. & Jonassen, D.H. (Eds), *Constructivism and the technology of instruction: a conversation.* Hillsdale, N.J.: Lawrence Erlbaum Associates.

De Bono, E. (1977). *Lateral thinking: a textbook of creativity.* Harmondsworth: Penguin Books.

De Villiers, M.R. (1999). Applying the Hexa-C Metamodel of instructional theory and design to educational web applications. In: de Bra, P. and Leggett, J. (Eds), *Proceedings of Web Net 99 – World Conference on the WWW and Internet.* Honolulu: Association for the Advancement of Computing in Education (AACE).

De Villiers, M.R. (2003). Foundation for structure: a learning theory approach to instructional design for e-learning. *International Journal of Learning Vol 10: Proceedings of the Learning Conference 2003*: 595-607. Melbourne: Common Ground Publishing.

De Villiers, M.R. (2005). E-learning artefacts: are they based on learning theory? *Alternation 12.1b*: 345-371.

De Villiers, M.R. & Cronjé, J.C. (2005). Six learning theory perspectives on a Web-based learning environment. *South African Journal of Higher Education, 19*(3): 444–466.

De Villiers, M.R. & Queiros, D.R. (2003). Real-world problem-based learning: a case study evaluated. *South African Journal of Higher Education, 17*(1): 112-122.

Dick, W. (1995). Instructional design and creativity: a response to the critics. *Educational Technology, 35*(4):5-11.

Dick, W. & Carey, L. (1996). *The systematic design of instruction* (4th Ed.). New York: Longman.

Dix, A., Finlay, J., Abowd, G.D. & Beale, R. (2004). *Human-computer interaction* (3rd ed.). Harlow: Pearson Education, Ltd.

Dodge, B. (1996). *Active learning on the Web.* Retrieved September 2006, from http://edweb.sdsu.edu/people/bdodge/Active/ActiveLearning.html

Duchastel, P. (1998). *Prolegomena to a theory of instructional design. ITForum Paper #27.* Retrieved September 2006, from http://it.coe.uga.edu/itforum/paper27/paper27.html

Duffy, T.M. & Jonassen, D.H. (1991). Constructivism: new implications for instructional technology? *Educational Technology, 31*(5):7-12.

Firdyiwek, Y. (1999). Web-based courseware tools: where is the pedagogy? *Educational Technology, 39*(1):29-34.

Gagné, R.M. & Glaser, R. (1987). Foundations in learning research. In: Gagné, R.M. (Ed.), *Instructional technology: foundations.* Hillsdale, N.J.: Lawrence Erlbaum Associates.

Gagné, R.M. & Merrill, M.D. (1990). Integrative goals for instructional design. *Educational Technology Research & Development, 38*(1):23-30.

Greeno, J.G., Collins, A.M. & Resnick, L.B. (1996). Cognition and learning. In: Berliner, D.C. & Calfee, R.C. (Eds), *Handbook of educational psychology.* New York: Macmillan.

Hannafin, M.J. (1992). Emerging technologies, ISD, and learning environments: critical perspectives. *Educational Technology Research & Development, 40*(1):49-63.

Hannafin, M.J., Hall, C., Land, S. & Hill, J. (1994). Learning in open-ended environments: assumptions, methods, and implications. *Educational Technology, 34*(8):48-55.

Hannafin, M.J., Hannafin, K.M., Land, S.M. & Oliver, K. (1997). Grounded practice and the design of constructivist learning environments. *Educational Technology Research & Development, 45*(3):101-117.

Hannafin, M., Land, S. & Oliver, K. (1999). Open learning environments: foundations, methods, and models. In: Reigeluth, C.M. (Ed.), *Instructional-design theories and models Volume II: a new paradigm of instructional theory.* Mahwah, N.J.: Lawrence Erlbaum Associates.

Hedberg, J., Brown, C, & Arrighi, M. (1997). Interactive multimedia and web-based learning: Similarities and differences. In Kahn, B.H. (Ed.), *Web-based instruction.* Englewood Cliffs, N.J: Educational Technology Publications.

Inhelder, B. & Piaget, J. (1958). *The growth of logical thinking from childhood to adolescence.* New York: Basic Books Inc. Publishers.

Johnson, D.W. & Johnson, R.T. (1991). *Learning together and alone.* Englewood Cliffs, N.J.: Prentice Hall.

Jonassen, D.H. (1992). Evaluating constructivistic learning. In: Duffy, T.M. & Jonassen, D.H. (Eds), *Constructivism and the technology of instruction: a conversation.* Hillsdale, N.J: Lawrence Erlbaum Associates.

Jonassen, D.H. (1994). Thinking technology: towards a constructivist design model. *Educational Technology, 34*(4):34-37.

Jonassen, D. (1999). Designing constructivist learning environments. In: Reigeluth, C.M. (Ed.), *Instructional-design theories and models Volume II: a new paradigm of instructional theory.* Mahwah, N.J.: Lawrence Erlbaum Associates.

Jonassen, D.H., Campbell, J.P. & Davidson, M.E. (1994). Learning with media: restructuring the debate. *Educational Technology Research & Development, 42*(2):31-39.

Jonassen, D.H. & Rohrer-Murphy, L. (1999). Activity theory as a framework for designing constructivist learning environments. *Educational Technology Research & Development, 47*(1):61-79.

Jones, M.G. (1998). *Creating engagement in computer-based learning environments. ITForum Paper #30.* Retrieved September 2006, from http://it.coe.uga.edu/itforum/paper30/paper30.html

Keller, J.M. & Suzuki, K. (1988). Use of the ARCS motivation model in courseware design. In: Jonassen, D.H. (Ed.), *Instructional designs for microcomputer courseware.* Hillsdale, N.J.: Lawrence Erlbaum Associates.

Land, S.M. & Greene, B.A. (2000). Project-based learning with the World Wide Web: a qualitative study of resource integration. *Educational Technology Research & Development, 48*(1):45-68.

Landa, L.N. (1998). Methodological drawbacks of current educational software and suggestions for radical improvement based on algo-heuristic theory. *Educational Technology, 38*(5):54-56.

Lebow, D. (1993). Constructivist values for instructional systems design: five principles toward a new mindset. *Educational Technology Research & Development, 41*(3):4-16.

Malone, T.W. (1981). Toward a theory of intrinsically motivating instruction. *Cognitive Science, 5* (but printed in footer as *4*):333-369.

Mayer, R.E. (1999). Designing instruction for constructivist learning. In: Reigeluth, C.M. (Ed.), *Instructional-design theories and models Volume II: a new paradigm of instructional theory*. Mahwah, N.J.: Lawrence Erlbaum Associates.

Merrill, M.D. (1983). Component display theory. In: Reigeluth, C.M. (Ed.), *Instructional design theories and models: an overview of their current status*. Hillsdale, N.J: Lawrence Erlbaum Associates.

Merrill, M.D. (2001). Components of instruction toward a theoretical tool for instructional design. *Instructional Science 29*: 291-310.

Minsky, M.L. (1975). A framework for representing knowledge. In: Winston, P.H. (Ed.), *The psychology of computer vision.* New York: McGraw-Hill.

Nelson, L.M. (1999). Collaborative problem solving. In: Reigeluth, C.M. (Ed.), *Instructional-design theories and models Volume II: A new paradigm of instructional theory*. Mahwah, N.J: Lawrence Erlbaum Associates.

Newell, A. & Simon, H.A. (1972). *Human problem solving.* Englewood Cliffs, N.J: Prentice-Hall Inc.

Norman, D.A. & Spohrer, J.C. (1996). *Learner-centered education. ITForum Paper #12.* Retrieved September 2006, from http://it.coe.uga.edu/itforum/paper12/paper12.html

Osman, M.E. & Hannafin, M.J. (1992). Metacognition research and theory: analysis and implications for instructional design. *Educational Technology Research & Development, 40*(2):83-99.

Panitz, T. (1996). *A definition of collaborative versus cooperative learning.* Retrieved September 2006, from http://www.lgu.ac.uk/deliberations/collab.learning/panitz2.html

Papert, S. (1980). *Mindstorms: children, computers and powerful ideas.* New York: Basic Books.

Perkins, D.N. (1991). Technology meets constructivism: do they make a marriage? *Educational Technology, 31*(5):18-22.

Price, E.A. (1998). Instructional systems design and the affective domain. *Educational Technology, 38*(6):17-28.

Preece, J., Rogers, Y., & Sharp, H. (2002). *Interaction design: beyond human-computer interaction.* New York: John Wiley & Sons.

Reeves, T.C. & Reeves, P.M. 1997. Effective dimensions of interactive learning on the World Wide Web. In: Khan. B.H. (Ed.). *Web-Based Instruction.* Englewood Cliffs, N.J: Educational Technology Publications.

Reigeluth, C.M. (1999). What is instructional-design theory and how is it changing? In: Reigeluth, C.M. (Ed.), *Instructional-design theories and models Volume II: A new paradigm of instructional theory.* Mahwah, N.J: Lawrence Erlbaum Associates.

Reigeluth, C.M. & Moore, J. (1999). Cognitive education and the cognitive domain. In: Reigeluth, C.M. (Ed.), *Instructional-design theories and models Volume II: A new paradigm of instructional theory.* Mahwah, N.J: Lawrence Erlbaum Associates.

Savery, J.R. & Duffy, T.M. (1995). Problem based learning: an instructional model and its constructivist framework. *Educational Technology, 35*(5):31-38.

Schön, D.A. (1987). *Educating the reflective practitioner.* San Francisco: Jossey-Bass Publishers.

Shneiderman, B. (1998). *Designing the user interface: strategies for effective human-computer interaction.* Reading, MA: Addison Wesley Longman.

Shneiderman, B., Borkowski, E.Y., Alavi, M., & Norman, K. (1998). Emergent patterns of teaching/learning in electronic classrooms. *Educational Technology Research & Development 46* (4): 23-42.

Simon, H.A. (1981). *The sciences of the artificial.* (2nd ed.). Cambridge, MA: MIT Press.

Singhanayok, C. & Hooper, S. (1998). The effects of cooperative learning and learner control on students' achievement, option selections, and attitudes. *Educational Technology Research & Development, 46*(2):17-33.

Smith, A.G. (1997). Testing the surf: criteria for evaluating Internet information resources. *The Public-Access Computer Systems Review, 8*(3). Retrieved June 2006, from http://info.lib.uh.edu/pr/v8/n3/smit8n3.html

Squires, D. (1997). *An heuristic approach to the evaluation of multimedia software.* Retrieved December 2005, from http://www.media.uwe.ac.uk/masoud/cal-97/papers/squires.htm

Squires, D. & Preece, J. (1999). Predicting quality in educational software: evaluating for learning, usability and the synergy between them. *Interacting with Computers 11*(5): 467–483.

Wager, W. (1998). Social determinants of affective behavior and learning. *Educational Technology, 38*(6):15-16.

West, C.K., Farmer, J.A. & Wolff, P.M. (1991). *Instructional design: implications from cognitive science.* Englewood Cliffs, N.J.: Prentice Hall.

Willis, J. (2000). The maturing of constructivist instructional design: some basic principles that can guide practice. *Educational Technology, 40*(1):5-16.

Willis, J. & Wright, K.E. (2000). A general set of procedures for constructivist instructional design: The New R2D2 Model. *Educational Technology, 40*(2):5-20.

Winn, W. (1990). Some implications of cognitive theory for instructional design. *Instructional Science, 19*(1):53-69.

Winn, W. (1992). The assumptions of constructivism and instructional design. In: Duffy, T.M. & Jonassen, D.H. (Eds.), *Constructivism and the technology of instruction: a conversation.* Hillsdale, N.J: Lawrence Erlbaum Associates.

Winn, W. (1997). Advantages of a theory-based curriculum in instructional technology. *Educational Technology, 37(*1): 34-41.

Winn, W. (1999). Learning in virtual environments: a theoretical framework and considerations for design. *Educational Media International, 36*(4): 271-279.

Author

Ruth de Villiers is a professor in the School of Computing of the University of South Africa. She teaches courses in human-computer interaction, e-learning and theoretical computing. Her cross-discipline research areas – in which she also supervisors postgraduate students – are design, usability and evaluation of educational technology applications; e-learning in general; HCI education; and meta-research.

Buzzetto-More, N. A. (Ed.). (2007). *Advanced Principles of Effective e-Learning*
Santa Rosa, California: Informing Science Press. (pp. 27-58)

Chapter 2

Using an Understanding of the Learning Cycle to Build Effective E-learning

Said Hadjerrouit

Introduction

E-learning is not a mode of education; it is a means of applying pedagogical theories to achieve better learning (Nocols, 2003). When implementing e-learning it is very important to be clear about the underlying pedagogical assumptions. A model of e-learning needs to demonstrate what pedagogical principles and learning theories are operating. Clearly, unless pedagogical principles are given more attention, e-learning practice cannot develop fully. Unfortunately, much of the development of e-learning is still carried out without a true understanding of how learning theories can be converted into pedagogical requirements (Govindasamy, 2002; Hamid, 2002; Nocols, 2003). In addition, e-learning lacks a systematic approach to the development process, resulting in poor analysis of pedagogical requirements that are proper to e-learning. Thus, to realize the potentialities of e-learning, it may be necessary to rely on a software development approach to translate learning theories into a system that provides learners with information that supports effective learning.

There exit many e-learning models and examples that are associated with learning theories (Collis & Moonen, 2001; Goodyear, 2001; ISM Global Learning Consortium 2002; Mayes & Fowler, 2005; Salmon, 2002; Scardamalia & Bereiter, 1994; Scott, 2005; Wiley, 2000), but few of them are pure derivatives of the three learning theories that will be discussed in this chapter: behaviorist, constructivist, and socially situated learning. In addition, few of them rely on a systematic software development approach from analysis to evolution.

The goal of this chapter is to inform teachers, learners, administrators, instructional designers, educational researchers and other stakeholders about how learning theories can be converted into pedagogical requirements that can be implemented using e-learning technologies. Mayes & Freitas (2005) argued that achieving better learning requires the understanding of pedagogical principles that are assumed in e-learning.

The remainder of this chapter is organized as follows. First, key terms are defined. Second, the main contemporary learning theories and associated e-learning cycle are described. Third, an e-learning development process that translates learning theories into a software system is specified. The process starts with mapping learning theories onto pedagogical requirements using the didactical relation model. This is followed by technical and pedagogical usability requirements, pedagogical modeling, pedagogical design, implementation, delivery and use, pedagogical evaluation, and finally, evolution. Fourth, the approach is implemented using the LMS Classfronter. Finally, the results of the work and some remarks on further work conclude the chapter.

E-learning

There is no clear and unequivocal definition of the concept of e-learning. Definitions in the research literature are partially exclusive and sometimes contradictory, and there are few common terms used consistently (Anohina, 2005; Nocols, 2003). It is difficult to distinguish the term "e-learning" from terms such as "virtual learning", "technology-based learning", "distance learning", "network learning", "online learning", "multimedia-based learning", "Web-enhanced learning", "Internet-enabled learning", and similar terms, because these are often used as synonyms.

E-learning is often seen as learning where the Internet and the Web play an important role. The term is also employed in a broader sense, namely as learning where any electronic technology is used, but it excludes aspects that might fit under "distance learning", but are not electronic, such as books. An attempt to define e-learning is to look at the relationships between e-learning and some closely related concepts: Internet, Web, online learning, and computer–based technologies.

First, the concept of Internet-based learning is broader than Web-based learning. The Web is only one of the Internet services that use

HTML, browsers, and URL. Internet offers many other services, not only Web, but also e-mail, file transfer facilities, etc. Learning could be based on the Web, but also as correspondence via e-mail.

Second, online learning could be organized through any network. Internet-based learning is only a subset of online learning.

Third, learning may take place via any electronic medium. It is not automatically connected to a network. Learning includes computer-based learning that is not network-based.

Finally, the term is employed in a broader sense, namely as learning that takes place via a combination of face-to-face and e-learning (Frank, Kurtz, & Levin, 2002). This kind of learning is called hybrid e-learning. It tries to overcome the disadvantages of pure e-learning.

To sum up, e-learning includes both network-based (online learning, Internet-based learning, and Web-based learning) and non-network-based learning (computer-based learning). Hybrid e-learning is a combination of e-learning and face-to-face learning (Figure 1).

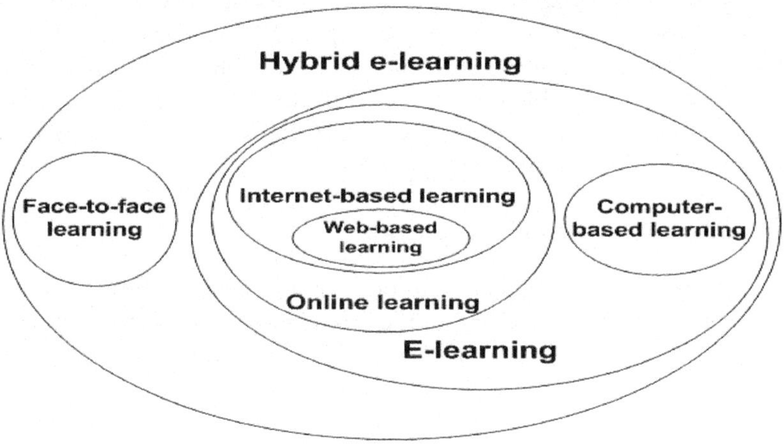

Figure 1. Components of e-learning and hybrid e-learning

Learning Theories

Important to the design of e-learning is a pedagogical foundation built on solid learning theory. Literature reviews suggest that learning theories can be related to three widespread models: behaviorist, constructivist, and socially situated model of learning. The remainder of this section describes the most important characteristics of the learning theories and presents a three-stage model - the learning cycle - that retains the features of each one.

The Behaviorist Perspective

The behaviorist perspective encompasses research traditions of associationism, behaviorism and connectionism. Learning is viewed as "the process of connecting elementary mental or behavioral units, through sequences of activity" (Mayes & Freitas, 2005). In terms of instruction, the behaviorist learning theory assumes that the goal of learning is to efficiently transmit knowledge from the instructor to the learners (Gagné, 1985; Skinner, 1976). In a behaviorist setting, instructors are clearly central to learning activities. However, while behaviorism promotes stability and certainty with respect to knowledge acquisition and learning outcomes, there are few opportunities for learners to express their owns ideas or concerns during the course of instruction, because behaviorism does not engage the mind appropriately to go beyond inappropriate prior knowledge. The behaviorist model is therefore criticized for stimulating surface learning and knowledge reproduction. On the other hand, behaviorist learning is suitable for novice learners, as they need transferable knowledge from the instructor. The behaviorist learning theory can be applied to e-learning to support the transmission of knowledge from the instructor to the learners.

The Constructivist Perspective

The constructivist perspective can be seen as a general shift in theoretical positioning in psychology and learning theory starting in the 1960s (Mayes & Freitas, 2005). In contrast to the behaviorist perspective, the constructivist paradigm views knowledge as a constructed entity made by each and every learner through a learning process. Constructivism frames learning less as the product of passive transmission than a process of active construction whereby the learners construct their own knowledge based upon prior knowledge and experience (Duffy, Lowyck & Jonassen, 1993; Piaget, 1971; Steffe & Gale, 1995). Constructivist learning requires learners to demonstrate their skills by constructing their own knowledge when solving real-world problems. Therefore, the constructivist model calls for learner-centered instruction because learners are assumed to learn better when they are forced to discover things themselves. In a constructivist setting, teachers serve primarily as guides and facilitators of learning, not as transmitters of knowledge. Accordingly, the constructivist learning theory can be applied to e-learning to support task-based activities rather than the transmission of knowledge from the instructor to the learners.

The Socially Situated Perspective

This perspective can be seen as a correction to constructivism, in which learning is disconnected from the social context. Whereas in the constructivist paradigm learning is assumed to occur as an individual learner interacts with study material, this perspective regards learning as socially situated and knowledge as socially distributed (Mayes & Freitas, 2005; Vygotsky, 1978; Wengler, 1998). Learning occurs as learners exercise, test, and improve their knowledge through discussion, dialogue, communication, collaboration, information sharing, and interaction with others. This means that knowledge is created as it is shared, and the more it is shared, the more it is learned. Vygotsky argued that the way learners construct knowledge, think, reason, and reflect on is uniquely shaped by their relationships with others. He argued that the guidance given by more capable others, allows the learner to engage in levels of activity that could not be managed alone. This guidance occurs in the Zone of Proximal Development (ZPD), which is the difference between what a learner can do independently and what can be accomplished cognitively with scaffolding from more knowledgeable others. Accordingly, the socially situated perspective of learning can be applied to e-learning to support collaborative learning, dialogue and discussion with both the instructor and fellow learners.

The Learning Cycle

The literature on learning theories points to the fundamental philosophical differences between them (Lin & Hsieh, 2001). However, in practice, a mix of learning theories is being used. Indeed, instructional designers tend to believe that what works in a learning situation is a subtle combination of learning theories. Hence, instructional designers must allow circumstances surrounding the learning situation to help them decide which approach to learning is most appropriate. It is necessary to realize that some learning problems require prescriptive/behaviorist solutions, whereas other are more suited to constructivist or situated/collaborative learning (Karagiorgi & Symeou, 2005).

Along the same line of argument, Mayes and Fowler (Mayes & Fowler, 1999) proposed a three-stage model or learning cycle, in which they identified three types of learning – conceptualization, construction, and dialogue. The essential characteristic of the learning cycle is that it describes a continuous cycle, or feedback loop, of gradual refinement of understanding. Accordingly, learning develops in three phases, beginning with conceptualization, progressing through construction to dialogue. Conceptualization is characterized by the process of interaction between the learners' pre-existing framework and instructor's knowledge. The construction phase – the intermediate phase of learning – refers to the process of building and combining concepts through their use in the performance of meaningful tasks. The dialogue phase refers to the testing of conceptualizations and the creation of new concepts during conversation with both fellow learners and instructors. Dialogue emerges through collaborative learning.

The three stages of the learning cycle include elements that are closely related to learning theories. Conceptualization is associated with the behaviorist learning theory as it focuses on the transmission of the instructor's knowledge to the learners. The construction phase is related to the constructivist learning theory as it aims at the construction of new knowledge through connecting and combining concepts, and their use in the performance of task-based activities. The dialogue phase is based on the socially situated learning theory as it is concerned with dialogue, group collaboration, and discussion.

At any stage of e-learning implementation there needs to be an emphasis on the three stages of the learning cycle: learning as behavior, learning as knowledge construction, and learning as dialogue and social practice. In this perspective it is important to consider the nature of the learning outcomes that are sought through the implementation of the learning cycle. Bloom's taxonomy of Educational Objectives (Bloom, 1956) is used as a general system for classifying learning outcomes and associated skills and competences to be achieved: knowledge, comprehension, application, analysis, synthesis, and evaluation. The conceptualization phase focuses on the acquisition of knowledge. The construction phase emphasizes comprehension, application, and analysis. The dialogue phase encourages synthesis and evaluation.

Information Technologies Used in the Learning Cycle

Mayes and Fowler characterized the types of information technologies used to achieve each stage of the learning cycle as primary, secondary, and tertiary courseware. Primary courseware is intended mainly to present the subject matter. Secondary courseware focuses on the set of software tools that support the performance of task-based activities. Tertiary courseware is the learning material which has been produced by previous learners. It may consist of dialogues between learners and teachers, as well as group discussions and collaborations. Mayes and Fowler's model has been adapted by Roberts (2003) to categorize three uses of the Web. Similarly, I have adapted this model to categorize three uses of e-learning (Figure 2).

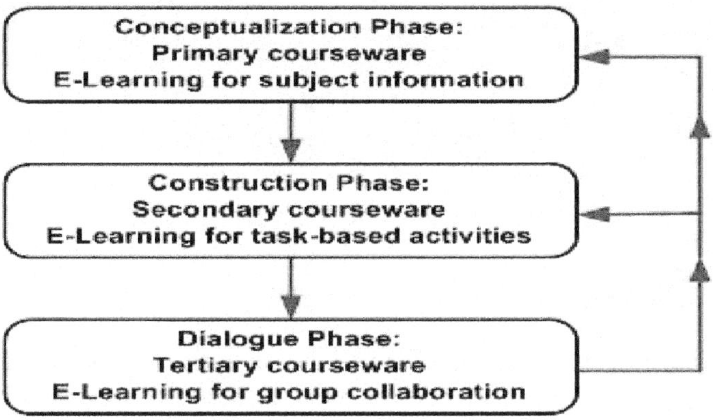

Figure 2. Mayes and Fowlers' learning cycle as feedback loop adapted to e-learning

E-Learning Development Process

For translating learning theories into systems that support effective learning, I propose an evolutionary process model. E-learning needs to evolve rapidly in order to ensure the relevance, correctness, and completeness of the content available online. A continuous, incremental and evolutionary development process is of crucial importance for ensuring the quality of e-learning. The development process is related to the activities that are required to produce e-learning. As illustrated in Figure 3, the development of e-learning is an orderly set of activities conducted and managed for each project. The development process identifies the phases along with e-learning moves from didactical analysis to evolution. The process includes eight phases.

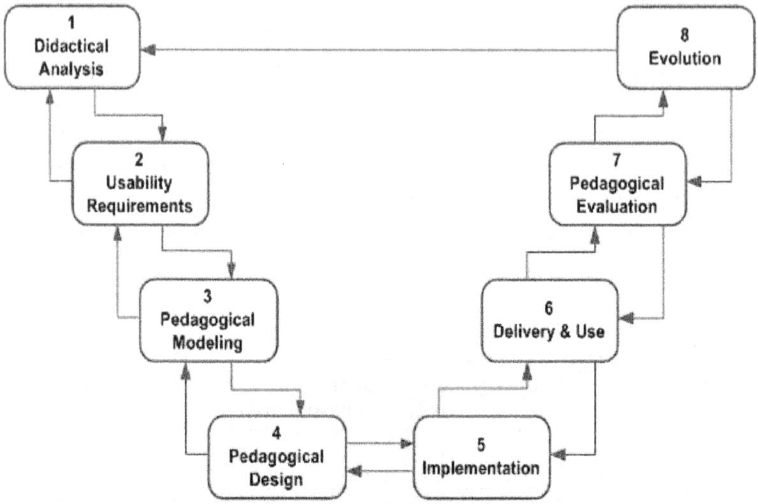

Figure 3. E-learning development process

Didactical Analysis

When applying learning theories to e-learning, it may be necessary to map the theories onto didactic categories. Such a mapping is a necessary step in any attempt to examine an e-learning implementation. Mapping onto didactic categories helps developers to bridge the gap between the theories and the technology. The major advantage of adopting such an approach is that it transposes learning theories to a pedagogical model well understood by teachers and instructional designers. For mapping learning theories onto didactic categories, there is a need for a didactical model that describes the broad principles that can be applied to the design of learning and teaching.

Didactical Relation Model

A didactical model describes the environment where teaching and learning take place. It is a space where learners can draw upon educational resources and teacher guidance to achieve meaningful learning.

The environment can be described using the didactical relation model (Hiim & Hippe, 1998). The environment assumes that different didactical elements are related to each other, and that there is a reciprocal influence between those elements. The didactical relation model consists of six dimensions: Learning outcomes, learners' abilities, structures and resources, learning content, teaching and learning methods, and assessment methods (Figure 4):

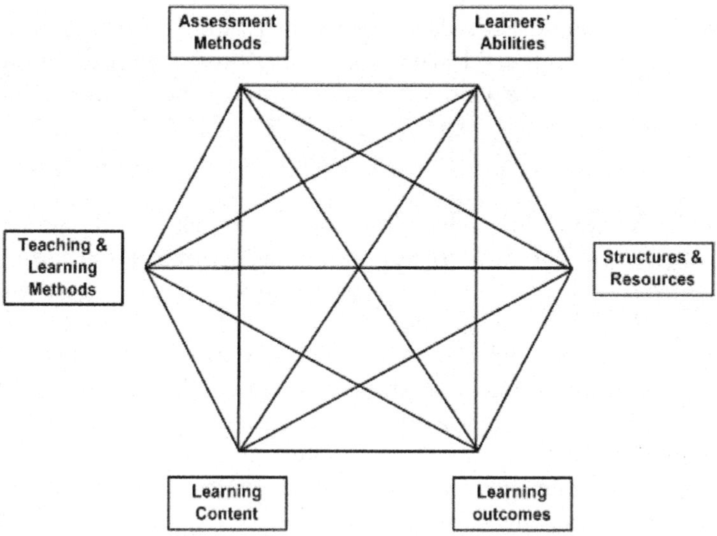

Figure 4. Teaching and learning environment based on the didactical relation model

- Learning outcomes are what the learners should possess after finishing learning activities in terms of concepts, methods, theory, practices, ideas, and principles.
- Learners' abilities are learners' prerequisite knowledge and skills, educational background, as well as personal experiences.
- Structures and resources are factors that influence education, such as technical and computer equipment, educational staff, library, books, time table, location, classroom settings, eco-

nomical conditions, legal and ethical conventions, curriculum, syllabus, etc.

- Learning content is the learning material that is associated with the subject matter, its topics and subtopics, and how these are broken down into lessons.
- Teaching and learning methods are concerned with activities and ways of working, such as lecturing, scaffolding, reading textbooks, doing exercises, performing projects, discussing and collaborating, etc.
- Assessment methods are concerned with the process of assessing the students' learning using assessment procedures such as oral and written exams, writing a report, performing a project, etc.

The didactical relation model helps teachers to plan and design courses according to pedagogical criteria. There are many ways of using this model. From a didactical point of view, teachers need to start with the intended learning outcomes, and then with the structures and resources. Alternatively, one could start with the specification of structures and resources in order to understand the external conditions that may influence learning, including learners' abilities. Then, they may define the learning outcomes. Furthermore, teachers need to specify the learning content and associated teaching and learning methods that can help students to achieve the learning outcomes. Finally, teachers must design assessment procedures in order to evaluate the students' learning in accordance with the learning outcomes. I propose the following planning sequence (Figure 5) for designing the teaching and learning environment according to the didactical relation model: (1) Analyze structures and resources; (2) Describe students' abilities; (3) Define learning outcomes; (4) Design learning content; (5) Specify teaching and learning methods; and (6) Describe assessment methods.

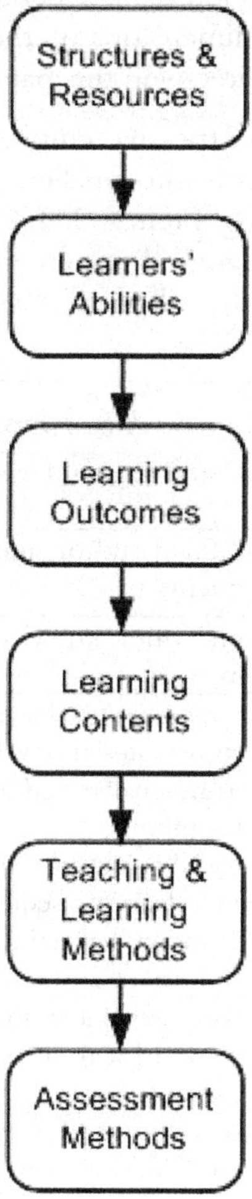

Figure 5. Course planning sequence using the didactical relation model

Mapping Learning Theories to Didactic Categories

Given the didactical relation model it is now possible to map learning theories onto didactic categories. The goal is to specify general guidelines that incorporate various aspects of the learning theories into a set of pedagogical requirements that may be implemented using e-learning technologies. It is worth acknowledging that some of the decisions that are taken in e-learning development depend on institutional structures and resources (one of the dimensions of the didactical relation model) that will not be addressed directly in this paper.

Basically, the behaviorist perspective emphasizes task analysis and decomposition of complex tasks into smaller components that are easy to learn (Gagne, 1985; Mayes & Freitas, 2005). It provides highly focused set of objectives. The pedagogical requirements that I derive from the behaviorist perspective are described in Table 1.

Table 1. Pedagogical requirements derived from the behaviorist learning theory

Didactic Categories	Features and Characteristics
Learning outcomes	▪ Well-defined and organized knowledge units and skill components
Learners' abilities	▪ Learners' prior knowledge and skills in the subject domain
Teaching & learning methods	▪ Use a transmission-based didactic mode of teaching, e.g. lecture notes and well-organized exercises ▪ Start from smaller and logically ordered units to more complex units ▪ Use a well-structured instructional method for each unit in the learning sequence ▪ Provide individualized instruction and routines ▪ Allow questions and immediate feedback
Learning contents	▪ Focus on the subject domain and its components ▪ Analyze and break down the domain into a hierarchy of small units ▪ Define clear and well-structured course topics, subtopics, and exercises for each unit ▪ Sequence the units logically and design an instructional approach for each unit ▪ Provide a well-structured and organized presentation of the subject domain
Assessment methods	▪ Summative assessment ▪ Assessment of well-defined knowledge units and skill components ▪ Written and oral tests, standard questionnaires

The constructivist perspective emphasizes conceptual understanding and stresses the importance of achieving understanding of the underlying principles of the subject matter. It focuses on the acquisition of cognitive skills, such as analysis reasoning, critical thinking, refection and evaluation (Duffy, Lowyck, & Jonassen, 1993; Mayes & Freitas, 2005). The pedagogical requirements that I derive from the constructivist perspective are described in Table 2.

Table 2. Pedagogical requirements derived from the constructivist learning theory

Didactic Categories	Features and Characteristics
Learning outcomes	▪ Critical thinking, reflection, and evaluation ▪ Conceptual understanding ▪ Analysis reasoning and analogical thinking
Learners' abilities	▪ Learners' prior knowledge and skills in problem-based and project-based activities ▪ Learners' individual characteristics and learning styles
Teaching & learning methods	▪ Provide support for active learning through questioning, critical thinking, and reflection ▪ Encourage experimentation, hypothesis building, inquiry and discovery of principles ▪ Foster problem-based and project-based learning, cognitive apprenticeships, and goal-based scenarios ▪ Provide teacher guidance and coaching ▪ Provide interactive environments for knowledge construction, exploration, and evaluation
Learning contents	▪ Concepts, ideas, and principles ▪ Authentic task-based activities taken from real-world situations ▪ Well-structured examples and case studies that learners may follow when they solve task-based activities and real-world problems ▪ Knowledge from previous versions of the course that learners may reuse with slight modifications ▪ Flexible learning material and multiple representation of information
Assessment Methods	▪ Formative assessment spread over the duration of the learning process ▪ Assessment of conceptual understanding and critical thinking

The socially situated perspective aggregates at the level of groups of learners, in which individuals participate as members of communities (Mayes & Freitas, 2005; Vygotsky, 1978; Wengler, 1998). The pedagogical requirements that I derive from the socially situated perspective are described in Table 3.

Table 3. Pedagogical requirements derived from the socially situated learning theory

Didactic Categories	Features and Characteristics
Learning outcomes	▪ Collaborative learning, dialogue and active participation in group work
Learners' abilities	▪ Learners' prior knowledge and skills in collaborative activities ▪ Learners' individual characteristics and learning styles
Teaching & learning methods	▪ Focus on group tasks and work ▪ Dialogue, collaboration, and social interaction ▪ Discussion and active participation among participants ▪ Feedback from instructor to the learners towards solutions to authentic problems ▪ Discussion groups, debates, role plays, brainstorming, project groups
Learning contents	▪ Discussions of research articles to reflect on the underlying ideas ▪ Participation and collaboration in real-world applications and projects to produce a solution to a problem ▪ Participation in social practices ▪ Dialogue that facilitates the development of learning relationships ▪ Discussions of case studies, project reports and application examples ▪ Collaborative assignments and shared tasks
Learning assessment	▪ Formative assessment spread over the duration of the learning process ▪ Qualitative evaluation, e.g. survey questionnaires, group interviews, and peer assessment ▪ Assessment of participation and collaboration in group work

Usability Requirements

This phase is concerned with the usability requirements that apply to e-learning. Usability requirements are twofold: technical and pedagogical usability.

Technical usability involves techniques and methods for ensuring a trouble-free interaction with e-learning while pedagogical usability aims at supporting the learning process. Both aspects of usability are closely related to each other (Melis, Weber & Andres, 2003). The goal should be minimizing the learners' work resulting from the interaction with e-learning in order to free more resources for the learning process itself.

Technical usability requirements

Technical usability is related to how a system is convenient, practicable, and usable for the learners. More specifically, Nielsen's factors of Web usability (Nielsen, 2000) include page and content design. Page design is related to cross platform, speed of page access, and page linking. Content design depends on writing for scannability and media use.

Accordingly, I consider the following criteria as important for the design of the technical usability of e-learning. These are divided into generic usability criteria that are applicable to most educational software and usability criteria that are specific to Web-based e-learning:

- Effectiveness, efficiency, user satisfaction, ease-of-use, and ease-of-learning
- Site structure, cross platform, accessibility, screen appearance, interactivity, navigation and linking, and multimedia design.

Pedagogical usability requirements

The technical usability is a self-evident requirement, but it is not sufficient for designing usable e-learning. The usability concept must be extended to capture issues that are fundamental to learning. The criteria that influence the pedagogical usability of e-learning are those that are associated with learning theories. The starting point for defining pedagogical usability requirements is to split the learning process into three types of learning with respect to the learning cycle: a behaviorist, a constructivist, and a dialogue phase. The suitable combination of the requirements produces a pedagogical usable e-learning.

The most important pedagogical usability requirements that have to be considered for behaviorism are (Mayes & Freitas, 2005):

- Break down the subject domain into a hierarchy of small learning units and logically discrete instructional steps
- Provide a well-structured online presentation of the learning units and instructional steps
- Allow easy and user-friendly accessibility of the subject information and links to related study material
- Provide a well-structured description of the subject information using a clear and understandable language

From a constructivist point of view, the most important requirements that influence the design of e-learning are the consideration of learners' individual characteristics and prior knowledge, authentic and motivating tasks, flexibility of learning material, multiple representation of information, reuse of information, etc. Accordingly, e-learning can be designed to support the following tasks (Mayes & Freitas, 2005; Wilson, 1998):

- Online well-designed examples that learners may follow when they perform task-based activities
- Online presentation of knowledge from previous versions of the e-learning system that students may adapt and reuse with some modifications to solve new problems
- Authentic task-based activities taken from real-world situations and online learning material that is intrinsically motivating
- Multiple representation of information using various multimedia elements (text, graphics, sound, animations, etc.)
- Teacher guidance and online feedback to learners' work and online submissions
- Links to interactive educational software for knowledge construction (Virtual Reality, games, simulations, multimedia software, etc.)
- Links to online databases, online journals, software libraries, interest groups, etc.

From the socially situated learning perspective, the most important requirements for e-learning are the consideration of social interaction, collaboration, and dialogue with fellow learners, teachers, and other stakeholders (Frank, Kurtz & Levin, 2002; Mayes & Freitas, 2005). E-learning can be designed to support the following tasks:

- Synchronous communication (chats), in which participants can communicate with each other simultaneously in real-time
- Asynchronous communication (discussion forum, e-mail, messages), in which participants are separated by time and apace
- Teachers and students may have access to a joint whiteboard. What is written or drawn there is shared in real-time
- Learners can submit a project, individually or as a group. Teachers can comment, grade the projects, give feedback, receive reports and documents online
- Arena for collaborative assignments to produce a shared project report, or solution to a problem
- Shared workspaces containing resources to all students. Such workspaces would allow collaborative work on shared tasks
- Online spaces that can be tailored according to the situational needs. This includes learners having the rights to add, modify, customize, manage, and delete items themselves

Pedagogical Modeling

This phase is concerned with a rigorous modeling of the pedagogical usability requirements defined in the previous phase of the development process. This phase specifies the users of e-learning, the functions they perform and system constraints.

Users

Basically, the users of e-learning are teachers, learners, and administrators:

- Learners use e-learning in order to participate to the educational process. In fact, learners are the most important users, in the sense that e-learning is being used in order to satisfy their educational needs.

- Teachers (instructors, tutors, etc.) use e-learning in order to provide new study material, update the existing one, supervise, coach, assist, and evaluate the students' learning, participate in discussions, communicate and exchange personal messages, collect, assess, return deliverables, comment project work, review learning activities, etc.
- Administrators undertake the support of all the users of e-learning. They administrate security and access rights to the system, network operations, monitor and repair database connections and operations, as well as server problems, maintain the system, produce statistics, etc.

Functions

Functions include the definition and specification of the set of operations that e-learning must provide to the users. These can be divided into teaching, learning, and administrative operations.

- Teaching operations are: register teacher information, teacher login, update course content (display, create, change, and delete course content), review task-based activities, answer questions and requests, assess student learning, check courseware evaluations, etc.
- Learning operations are: register learner information, learner login, display course content, perform task-based activities, perform dialogue and group collaboration, perform project work, perform assessment, perform courseware evaluation, etc.
- Administrative operations are: update user information, produce statistics, and create security and access rights.

System constraints

System constraints describe how e-learning is constrained when accomplishing its functions (Maciaszek, 2001). Accordingly, e-learning constraints are set with regards to:

- Performance requirements
- Security requirements
- Operational requirements
- Political and legal requirements

Performance requirements can become quite central to the success of e-learning. They specify the speed (the system's response time) at which various tasks have to be accomplished. Performance requirements ensure a trouble free function of e-learning.

Security requirements describe user's access privileges to the information under the system's control. Users can be given restricted access to the e-learning, including restricted access to data and/or restricted rights to execute certain operations on data. Some of these requirements are related to users' requirements.

Operational requirements determine the hardware/software environment in which e-learning will operate. These requirements may have an impact on other aspects of the management process of e-learning, such as system maintenance and update.

Political requirements are frequently assumed rather than explicitly stated. These requirements are derived from the institutional environment, and specify the institutional, legal, and ethical issues. These requirements are very important because e-learning may be difficult or impossible to use for political, legal, and ethical reasons.

Pedagogical Design

I recommend e-learning that is hierarchical with the top-level as the main page presenting general information, login and registration procedures. The system is decomposed into five major components: conceptualization, construction, dialogue, learning assessment, and courseware evaluation. Each component is broken down into smaller components according to the requirements described in previous sections. Figure 6 gives an overview of the pedagogical design and the resulting e-learning model. Only the most important components are represented.

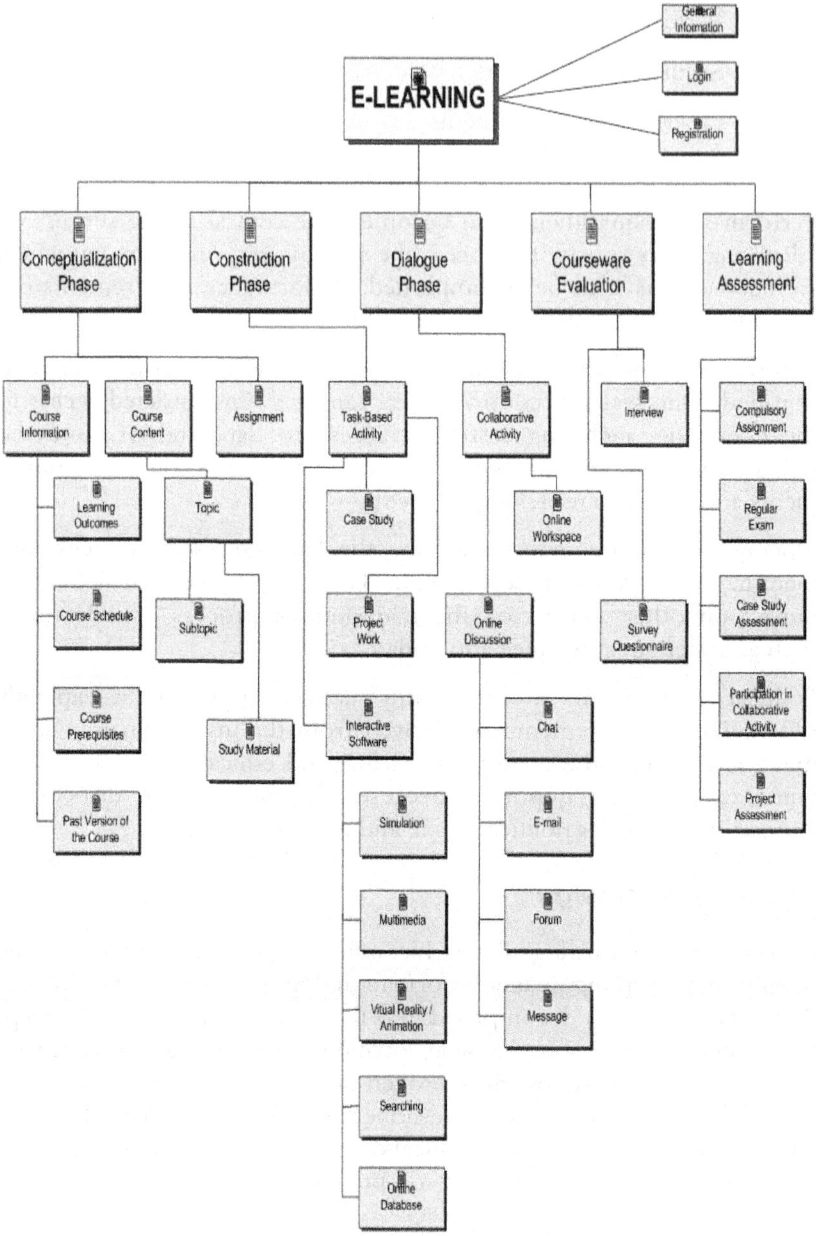

Figure 6. Recommended model of e-learning

Implementation

Basically, there are two ways of implementing the proposed e-learning model. First, it may be implemented from the ground using programming languages, specialized authoring systems, and tools such as FrontPage or Macromedia Dreamweaver for Web pages, or similar software tools. Second, the model may be implemented using standard software such as LMSs (Learning Management Systems) as these are the most common modes for delivery. In this case, it would not be necessary to implement the model from the ground since it is possible to use an existing LMS as a platform for e-learning. In addition, independently of the chosen solution, it is recommended to modify, refine, and reuse e-learning components (Krauss & Ally, 2005; Rokou, Rokou & Rokos, 2004). Components may include reusable course units, lessons, exercises, assignments, learning activities, project reports, case studies, past exams, evaluation questionnaires, students' discussions, links to interactive software, and other files, documents, multimedia elements, etc.

Before delivery, e-learning must be tested systematically. Testing is the process of exercising the system with the intent of finding and ultimately correcting various errors, such as typographical errors, grammatical mistakes, errors in content, errors in graphical and multimedia representations, cross referencing errors, navigation errors, etc.

Pedagogical Evaluation

The evaluation of e-learning can begin once it is used in authentic educational environments. The evaluation ensures that the objectives are kept in mind, and that the decisions made throughout the phases of usability requirements definition and modeling, design, implementation, and testing are achieved.

The evaluation must be conducted not only for the purpose of reviewing and revising the technical quality and the quality of the learning material but also for the evaluation of learning and the underlying pedagogy, since these are the driving forces behind the development of e-learning.

There are many methods and instruments that may be used to assess the pedagogical value of e-learning, for example online standard questionnaires, individual and group interviews, and eventually video-taped observation sessions, are relevant instruments to obtain data on what learners feel and think about e-learning (Dyson & Campello, 2003; Storey, Phillips, Maczewski & Wang, 2002; Williams, 2002).

In addition, learners' examinations and compulsory test performances are also adequate data for assessing the effectiveness of e-learning for acquiring skills and knowledge. Finally, literature reviews, teacher reflections and expert opinions in educational research are also good instruments to elicit the pedagogical value of e-learning.

Evolution

Considering that e-learning will undergo change once it is used and evaluated, it is necessary to plan an evolution phase in order to ensure that the content is updated and revised, and the pedagogy is improved. A continuous evolution is of crucial importance for the quality of e-learning.

Case Study Using the LMS Classfronter

To illustrate the application of the e-learning model an example is proposed, showing the steps teachers and students would go through in using it. The example is based on the course on Practical ICT Didactics taught in the academic year of 2005-2006 at Agder University College. The LMS Classfronter was used a platform for e-learning. The course relied on hybrid e-learning, a combination of face-to-face and e-learning.

LMS Classfronter

Classfronter is a Learning Management System (LMS that is a completely Web-based (Classfronter, 2006). It is based on a building metaphor. It may therefore be regarded as a huge virtual building containing corridors and rooms. The main purpose of corridors is to organize rooms, and they may not contain anything else. Rooms are therefore the main Classfronter object. They may contain both learning resources and activities. All Classfronter users have their own, private room, "My room". In addition, most users may access several shared rooms. A room for a class or a course is a typical example of shared rooms where students and instructors in that course have shared room access. This gives them a joint, electronic arena where they may share learning resources, collaborate and cooperate in different ways. More specifically,

Classfronter supports the development and execution of the following basic tasks:

- Students can work together, develop project reports by writing in the same document, exchange information, pass messages to each other, share ideas and opinions, create a portfolio of goals for their studies, etc.
- Teachers, students, and other participants have access to multiple communication facilities, e.g. synchronous dialog for communicating simultaneously in real-time, and asynchronous communication, in which participants are separated by time and space.
- Students can submit project, individually or as group. Teachers can comment or grade the projects, and obtain reports.
- Students and teachers in a room have access to a joint whiteboard, where they can write documents and draw figures. What is written or drawn there is shared in real-time.
- Teachers can publish messages to participants, publish lecture notes, minutes of meetings, presentations and other documents so these will always be available.
- Teachers can create tests and surveys which can be executed and corrected automatically in Classfronter. They can date information regarding participants, follow up the portfolio of the participants, create online marking of students' assessment, etc.

The E-learning Model in Use

The starting point for applying the e-learning model to LMS Classfronter was to split the learning process into three phases of learning according to the learning cycle: a conceptualization phase, a construction phase, and a dialogue phase:

First, Classfronter was designed to support the conceptualization phase - that is to say the process of interaction between the learners' preexisting knowledge and the key concepts of the subject matter. As a result, Classfronter was used as a source for subject information, enabling the access to resources that offer various types of information that can be used to either a greater understanding of the subject matter, or to obtain further information about it. The most important criteria for designing Classfronter for conceptualization were a well-structured presentation, easy accessibility, and powerful explanation of the information in order to effectively transmit knowledge to the learners.

Second, Classfronter was designed to support the construction phase – that is to say the process of building new conceptualizations through the performance of task-based activities. In order to perform their tasks, students should have access to resources that support active, independent, and self-reflective learning. These resources require the design of effective tasks - rather than the presentation of the subject matter - in order to encourage learners to think conceptually. As a result, the resources for the construction phase were designed to contain study material for task-based learning, e.g. well designed examples of case studies that may be reused with slight modifications, research-based project tasks, links to a wide range of educational software, pedagogical applications that may be adapted, modified, and extended to meet the requirements of students' work.

Third, Classfronter was designed to support the dialogue phase of the learning cycle, enabling students to test their project work through conversations, collaboration, and reflection with the instructor and follow students, to perform synchronous and asynchronous discussions of articles taken from international journals, to share workspaces containing resources to all students, to reuse dialog depositories from past versions of the course, to evaluate the courseware, etc.

Conceptualization Phase

Before the conceptualization phase took place, students had to understand how to use Classfronter and its main functionality: login and logout procedure, organization of rooms and corridors, access to shared rooms, the "Today" page of the rooms, message boards, students' portfolio and archives, contact register, calendar, help function, download of course material, upload file, discussion groups, chats, whiteboard, etc. The goal for the students was to learn to use Classfronter as a learning platform in the most efficient, effective manner possible. To do this, they had access to online documentation that presents the LMS in a clear and understandable manner.

Once the software was understood and sufficient knowledge about the functionality of the system was transmitted to the students, the focus shifted from the LMS to the conceptualization phase. The goal of this phase consisted of setting the context and the conditions of learning activities. Students had to rely on Classfronter for information at the beginning of any learning activity. The instructor directed student learning by presenting the subject matter through Classfronter. To develop clear educational objectives, the subject matter had to be broken down into its components.

Because behaviorism is governed by an objective view of the nature of knowledge, the transition from the teacher's information to the students was not entirely difficult. This involved an initial interaction with the subject matter presented through the LMS classfronter. This interaction produced an initial interpretation of the subject matter trough reading the course material and doing the exercises available online.

A first online evaluation was conducted after one month of using Classfronter. The evaluation was concerned with collecting data about e-learning with respect to the conceptualization phase of the learning cycle - that is to say the process of interaction between students' preexisting knowledge and the level of difficulty, scope, and depth of the subject matter, its topics and subtopics.

The first evaluation enabled students to suggest improvements and changes. Then, Classfronter was redesigned according to the students' recommendations. The goal was minimizing the students' work load resulting from the interaction with the LMS in order to help them to concentrate on what really matters: the learning of the subject matter.

Improvements consisted of improving the ease-of-use and learner satisfaction with Classfronter, e.g. reorganizing the folders and associated course material, reducing technical hurdles and frustration when interacting with the software. In addition, course material that was quite difficult for the students and not very important for the learning process was partly removed from Classfronter.

Construction Phase

The second phase of e-learning – the construction phase of the learning cycle - assumed that the students have understood the main functionality of the LMS so that they were able to use it in an efficient way to study the course content of the subject matter.

The constructivist phase refers to the process of building new knowledge through the performance of authentic tasks. The emphasis here is on supporting students to think about and reflect on the subject matter at a deeper conceptual level than in the conceptualization phase. For achieving this, students should have access to large sources of relevant information about the subject matter and motivating activities aimed at the construction of new knowledge. Classfronter was used in a variety of ways.

First, Classfronter provided links to a wide range of pedagogical software that may be used for exploration and evaluation (animations, simulations and multimedia, etc.). These applications included a wide range of school fields (mathematics, physics, etc.) at different levels.

Second, Classfronter provided well-structured examples of knowledge application. Students studied the examples, reused some of them, and wrote their own reports describing their experiences and submitted them electronically to the instructor. In order to document their reports with relevant literature, students needed to find and assemble relevant articles and documents.

Third, Classfronter offered well-designed case studies and research-based project reports that may be reused with slight modifications. The goal was to provide motivation, stimulate reflection, help reorganize students' thinking, and evaluate their work when performing research-based activities.

The information available in Classfronter was presented in multiple forms using various media elements and formats (Word documents, Web pages, Power Point, PDF, images, animations, etc.). The goal was to make the recording of learning material intrinsically motivating.

Since the constructivist learning theory recognizes that the human mind is not simply a passive recipient of knowledge, this phase required being clear about students' prior knowledge, skills, and learning styles before the construction phase came into play. The role of the instructor was to provide feedback whenever students asked questions.

Dialogue Phase

E-learning continued with the dialogue phase of learning. This phase referred to the testing of ideas, opinions, and solutions developed in the construction phase. The goal was to test and further develop the solutions during conversations and collaboration with both fellow students, and the instructor and in the reflection on these. Students used Classfronter in a variety of ways in this phase.

First, Classfronter was used as a forum for dialogue through LMS-enabled asynchronous discussion forum with the instructor and fellow students. The discussions were about research articles taken from international journals. The goal was to reflect on the content of the papers and the underlying ideas using a set of criteria, and then to write a summary of the articles and the reflection on theses. Students could explore dialogue material which had been produced by previous learners, in the course of discussing and assessing research articles. This material was available online. Asynchronous discussions gave the students the opportunity to concretize their ideas before responding. This often encouraged in-depth, more thoughtful discussion.

Second, students submitted their compulsory project work electronically and received the instructor's comments on their work electronically as well. The goal was to help the students improve their writings and associated project reports. Students benefited from reflecting on how they had achieved the learning goals, before they submitted their reports to the instructor. What difficulties were encountered? What caused the difficulties? How will they overcome them? Feedback and comments from the instructor gave students the necessary pieces of how to perform project work. This phase helped students to develop their own ability to self-assess their own project work.

Third, students together with the instructor built shared workspaces containing resources to all students. Such workspaces allowed collaborative work and peer reviews on shared tasks such as writing reports. This allowed Classfronter to be used as an arena for collaborative writing to produce shared tasks and reports.

Fourth, students used Classfronter to present the results of their work to the whole class. During the entire learning process, students were communicating the results of their efforts in an informal way and discussing what they had learned with other students and the instructor. To formally communicate their results, students presented their work using Classfronter, showing others what they had done and discussing what they learned.

Finally, a second evaluation was conducted and delivered electronically through Classfronter to the students one week before the end of the course. The evaluation used a questionnaire for collecting both quantitative and qualitative data. The questionnaire of the second evaluation contained more issues than the one used in the first evaluation: cognitive aspects, difficulties, learning problems, collaboration with the instructor, etc.

As a result, the goal of this phase was for the students to acquire specific skills that are of crucial importance from a socially situated learning perspective: discussing project work with the instructor, writing and formatting understandable documents for others, reading and reflecting on the writings of other participants, collaborating and interacting with other learners, reusing students' work and experiences from previous students, etc.

Discussion and Conclusion

The main goal of this chapter is to outline a model for translating learning theories into pedagogical requirements that can be applied when selecting and using e-learning technologies. The requirements have been derived from three theoretical perspectives: behaviorist, constructivist and socially situated learning, and the combination of these to the learning cycle with three stages corresponding to the learning theories. I framed this approach within the didactical relation model, which describes the educational environment where teaching and learning take place. The major advantage of adopting such an approach is that it transposes learning theories to a pedagogical model well understood by teachers. Given this background, the following conclusions can be drawn.

First, the learning cycle provides an adequate pedagogical framework for designing e-learning through a cyclical dynamic process starting from conceptualization, progressing through construction to dialogue. More specifically, the learning cycle provides an adequate support for the conceptualization phase - that is the interaction between the learners' prerequisite knowledge and the subject matter. Second, the learning cycle provides sufficient support for the construction phase of the learning cycle - that is the process of performing task-based activities. Finally, the dialogue phase of the learning cycle is also integrated into e-learning.

Second, even though it would be possible to improve the overall design of e-learning through a strict development process and rigorous analysis and pedagogical design, I believe that e-learning alone is not sufficient for implementing the dialogue phase of the learning cycle. In authentic educational settings, teachers do not just convey subject information and task-based activities that may be available online, but must act as a facilitators, guides, and mentors, where dialogue plays a central role. Dialogue can be supported by tertiary courseware through online discussions, both synchronous and asynchronous, but it is a human relation as well. However, online dialogue lacks the very basis of any human dialogue: face-to-face meetings, eye-contact, body language, and non-verbal behavior between the teacher and the students.

References

Anohina, A. (2005). Analysis of the Terminology used in the Field of Virtual Learning. *Educational Technology & Society*, 8 (3), 91-102.

Bloom, B. S. (1956). *Taxonomy of Educational Objectives: The Classification of Educational Goals*. New York: Longman.

Classfronter (2006). *Classfronter. Agder University College*. Retrieved November 25, 2005, from: https://www.hia.no/classfronter

Collis, B., & Moonen, J. (2001). *Flexible Learning in a Digital World*. London: Kogan Page.

Duffy, T.M., Lowyck, J., & Jonassen, D.H. (1993). *Designing Environments for Constructive Learning*. Berlin: Springer.

Dyson, M. C., & Campello, S. B. (2003). Evaluating Virtual Learning Environments: What are we measuring? *Electronic Journal of E-Learning*, Volume 1, Issue 1, 11-20.

Frank, M., Kurtz, G., & Levin, N. (2002). Implications of Presenting Pre-University Courses Using the Blended e-Learning Approach. *Educational Technology & Society* 5(4), 2002, 137-147.

Gagne, R. (1985). *The Conditions of Learning*. New York: Holt, Rinehart and Winston.

Goodyear, P. (2001). *Effective Networked Learning in Higher Education: Notes and Guidelines*. Retrieved March 23, 2006, from: http://www.csalt.lancs.ac.uk/jisc/guidelines_final.doc

Govindasamy, T. (2002). Successful Implementation of E-learning: Pedagogical Considerations. *The Internet and Higher Education*, 4, 287-299.

Hiim, H., & Hippe, E. (1998): *Instructor Guidance in the Education of Teachers related to Vocational Subjects* [In Norwegian: Undervisningsplanlegging for yrkeslærere]. Universitetsforlaget: Oslo.

Hamid, A. A. (2002). E-Learning: Is it the "E" or the Learning that Matters? *The Internet and Higher Education*, 4, 311-316.

Hasarim, L. (2000). Shift Happens: Online Learning as a New Paradigm in Learning. *The Internet and Higher Education* 3, 41-61.

ISM Global Learning Consortium (2002). *IMS Learning Design: Best Practice and Implementation*. Retrieved March 18, 2006, from: http://www.imsglobal.org/learningdesign/ldv1p0/imsld_bestv1p0.html

Karagiorgi, Y., & Symeou, L. (2005). Translating Constructivism into Instructional Design: Potential and Limitations. *Educational Technology & Society*, 8(1), 17-27.

Krauss, F., & Ally, M. (2005). A Study of the Design and Evaluation of a Learning Object and Implications for Content Development. *Interdisciplinary Journal of Knowledge and Learning Objects*, Vol. 1, 1-22.

Lin, B., & Hsieh, C. (2001). Web-based Teaching and Learner Control: A Research Review. *Computers & Education*, Volume 37, Issues 3-4, 377-386.

Lowyck, J., & Poeysae, J. (2001). Design of Collaborative Learning Environments. *Computer in Human Behavior*, Vol. 17, Issues 5-6, 507-516.

Maciaszek, L.A. (2001). *Requirements Analysis and System Design: Developing Information Systems with UML*. New York: Addison-Wesley.

Mayes, J. T., & Fowler, C. J. (2005).*Mapping Theory to Practice and Practice to Tool functionality based on the Practitioners' perspective*. Retrieved March 28, 2006, from: http://www.jisc.ac.uk/uploaded_documents/Stage%202%20Mapping%20(Version%201).pdf

Mayes, J. T., & Fowler, C. J. (1999). Learning Technology and Usability: A Framework for Understanding Courseware. *Interacting with Computers*, Volume 11, Issue 5, 485-497.

Mayes, T., & Freitas, S. (2005). *Review of e-Learning Theories, Frameworks and Models*. Retrieved March 23, 2006, from: http://www.jisc.ac.uk/uploaded_documents/Stage%202%20Learning%20Models%20(Version%201).pdf

Melis, E., Weber, M. & Andres, E. (2003). Lessons for (Pedagogic) Usability of eLearning Systems. *Proceedings* of *E-LEARN 2003*, Phoenix, Nov. 7-11, 281-284.

Nielen, J. (2000). *Designing Web Usability: The Practice of Simplicity*. New York: New Riders.

Nocols, M. (2003). A Theory of Learning. *Educational Technology & Society*, 6(2), 1-10.

Piaget, J. (1971). *Genetic Epistemology*. New York: W.W. Norton.

Roberts, G. (2003). Teaching Using the Web: Conceptions and Approaches from a Phenomenographic Perspective. *Instructional Science* 31, 127-150.

Rokou, F. P., Rokou, E., & Rokos, Y. (2004). Modeling Web-Based Educational Systems: Process Design Teaching Model. *Educational Technology & Society*, 7(1), 2004, 42-50.

Salmon, C. (2002). E-Tivities: The Key to Active Online Learning. London: Kogan Page.

Scardamalia, M., & Bereiter, C. (1994). Computer Support for Knowledge-Building Communities. *The Journal of the Learning Sciences*, 3(3), 265-283. Retrieved February 12, 2006, from: http://carbon.cudenver.edu/~bwilson/building.html

Scott, J. (2005). *Assessing the Relevance of the Review of E-learning Theories, Frameworks and Models and the Mapping Table to Designers*. Retrieved April 19, 2006, from: http://www.jisc.ac.uk/uploaded_documents/Designers.doc

Skinner, B. F. (1976). *About Behaviorism*. New York: Vintage.

Steffe, L. P., & Gale J. (1995) (Ed.). *Constructivism in Education*. New Jersey: Lawrence Erlbaum Associates.

Storey, M.A., Phillips, B., Maczewski, M. &. Wang, M. (2002). Evaluating the Usability of Web-Based Learning Tools. *Educational Technology & Society*, Volume 5, Issue 3, 91-100.

Vygotsky, L.S. (1978). *Mind in Society: The Development of Higher Psychological Processes*. Cambridge. MA: Harvard University Press.

Wengler, E. (1998). *Communities of Practice: Learning, Meaning, and Identity*. London: Pinter.

Wiley, D.A. (2000). *Connecting Learning Objects to Instructional Design Theory. A Definition, a Metaphor, and a Taxonomy*. Retrieved January 22, 2006, from: http://reusability.org/read/chapters/wiley.doc

Williams, D. D. (2002). Improving Use of Learning Technologies in Higher Education through Participant Oriented Evaluations. *Educational Technology & Society* 5(3), 11-17.

Wilson, B. G. (Ed.) (1998). *Constructivist Learning Environments: Case Studies in Instructional Design*. New Jersey: Educational Technologies Publications.

Author

Said Hadjerrouit received the MS and PhD degrees in Software Engineering and Artificial Intelligence from the Technical University of Berlin (Germany), in 1985 and 1992, respectively. He joined Agder University College, Kristiansand (Norway) in 1991. He is currently an Associate Professor of Computer Science at the Faculty of Mathematics and Sciences. He has been in the teaching profession for 24 years. He has extensive experience teaching object-oriented programming, Web design, database development, and software engineering. His research interests include computer science and software engineering education, didactics of information technology, E-Learning, and Web engineering education. E-mail: Said.Hadjerrouit@hia.no

Buzzetto-More, N. A. (Ed.). (2007). *Advanced Principles of Effective e-Learning*
Santa Rosa, California: Informing Science Press. (pp. 59 111)

Chapter 3

Concept Mapping as a Means to Build E-Learning

Christina M. Steiner, Dietrich Albert, and Jürgen Heller

Introduction

Concept maps constitute graphical node-link representations that specify the concepts of a knowledge domain and the relations among them (e.g. Coffey et al., 2003; Lanzing, 2004; Novak, 1998, 2001). Since the information concept maps represent is semantic knowledge in a general sense, they are sometimes also called semantic networks (Lanzing, 2004). Because of its multipurpose applicability, concept mapping and related mapping strategies have come to the fore in the last few decades. Particularly with the increase of computer support for learning and education, concept mapping has achieved more and more popularity. Concept maps provide useful and facilitative instruments in many stages of planning, developing and carrying out distance education. Concept mapping can be applied as a strategy for modeling an e-learning course. Course designers and content experts are assisted in eliciting and determining the respective knowledge and in systematizing it. On the one hand, a concept map can be used to model content based relations, specifying the relationships existing among the content. On the other hand, 'instructional relations', such as prerequisite or corequisite relations among concepts, can be captured by a concept map, providing information on the instructional sequence. By organizing contents in this way, a structure is provided for creating and sequencing units of learning. Concept maps that result from the planning process and depict content based relations can also be used for representing learning content in the course itself, e.g. by inserting media resources that are associated with concepts. Concept mapping can also be implemented into an e-learning system as an active learning strategy, i.e. individuals or groups of learners create, complete or correct concept maps during a unit of learning. Computerized knowledge assessment methods on the basis of concept maps are a proper option for evaluation of learning progress and outcome. This is just a short sketch of the numerous possibilities for effective and beneficial application of concept maps.

But why is concept mapping successful? The usefulness of and interest in concept mapping can be explained from the perspective of cognitive psychology. Concept mapping stands in the tradition of the networking theories of cognitive psychology. These theories arose in the 1960's and 70's and are grounded in the notion that long-term memory can be seen as an internal semantic network structure (Chiarello, 1994; Sternberg, 1999, chap. 8; Wender, 1988). Such theories postulating that there is a correspondence between graphical network representations and memory were proposed by Quillian (1966), Collins and Quillian (1969), Collins and Loftus (1975), Anderson and Bower (1974), Rumelhart, Lindsay, and Norman (1972, see also Lindsay & Norman, 1977). This nurtures the idea of representing an individual's personal knowledge by a network representation like a concept map. Because of the claimed structural similarity between graphical network representations and the mental representation of long-term memory, it is assumed that the use of such network structures for presenting and communicating information is beneficial (Mandl & Fischer, 2000). In sum, the fact that concept maps are assumed to be able to reflect semantic structures of human memory accounts for their widespread and effective use.

This chapter highlights the versatile areas of application for concept mapping in educational contexts, and in the scope of e-learning. It introduces the basics of concept maps, outlines how they are built in practice and how this process can be supported by computer-based tools. Focusing on pedagogical aspects it discusses the role of concept mapping in instructional planning and its potential for developing teaching and learning strategies for individuals as well as for groups. Moreover, it is shown that there is a variety of concept mapping techniques and tasks, which may be exploited for knowledge assessment. Finally, the chapter addresses the benefits that concept mapping offers for improving the design of e-learning systems and the navigation through them. In particular, it draws attention to the innovative development of adaptive e-learning systems based on prerequisite relations among learning objects and competencies that are induced by concept maps. The chapter concludes with an outlook on future work.

Definition and Representation of Concept Maps

Concept maps provide a tool for structuring and representing knowledge, depicting the concepts of a content area and the relationships that exist between them (e.g. Novak, 2001). Therefore, they provide a natural way of expressing and presenting ontologies of any knowledge domain.

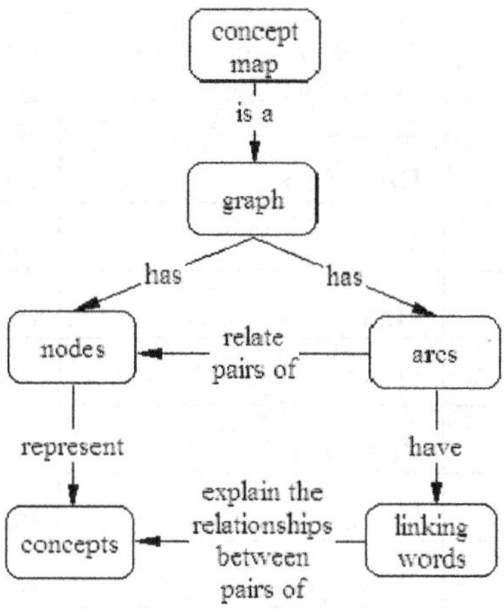

Figure 1: Concept map describing what a concept map is (adapted from Ruiz-Primo, 2000)

There are different types of representations of a concept map. Most commonly, a concept map is visualized through graphical representation. Concepts are usually depicted by circles or boxes, forming the nodes of the network. The relations among concepts are indicated and explained by labeled links (arcs). The directions of relations are usually indicated by arrowheads. Figure 1 shows an example of a concept map that represents what a concept map is (for further examples see also Figure 6).

The combination of two concepts and the link relating them forms a proposition. "A proposition is the smallest unit of knowledge that can stand as a separate assertion; that is, the smallest unit about which it makes sense to make the judgment true or false" (Anderson, 1995, p. 145). Another way of representing a concept map is therefore through a list or set of propositions. Table 1 represents the same concept map as shown in Figure 1, but in the form of a list of propositions. When generating such a proposition list by the use of a spreadsheet software (e.g. Microsoft Excel), such a list can easily be rearranged and sorted according to different characteristics – for instance with respect to the relation labels. Alternatively, a concept map may be represented in the form of a matrix with the set of concepts labeling the columns and rows and the relations specified in the cells of the matrix.

Table 1: Concept map describing what a concept map is, represented by a list of propositions

Concept	**Relation**	**Concept**
concept map	is a	graph
Graph	Has	nodes
Graph	Has	arcs
Nodes	Represent	concepts
Arcs	Have	linking words
Arcs	relate pairs of	nodes
linking words	explain the relationships between pairs of	concepts

How to Build a Concept Map

In general, a concept map can be constructed by an individual, but also as part of a collaborative group process (Coffey et al., 2003). Several authors tried to outline a step-by-step procedure for constructing a concept map (e.g. Chrobak, 2001; Novak & Gowin, 1984; Zimmaro & Cawley, 1998). When getting into contact with concept mapping for the first time and learning how to construct a concept map, a knowledge domain should be addressed that is familiar to the person constructing the map. Furthermore, the knowledge in question should be clearly defined and limited (Novak, 2001).

Constructing a concept map involves the following steps (adapted from Novak and Gowin, 1984):

- Defining a topic or focus question, for which the concept map is to be built.
- When the key topic has been defined, the most relevant concepts (key concepts) that are associated with this topic have to be collected and listed.
- Next, the concepts are arranged in the mapping field in a pattern that best represents the information.
- After ordering the key concepts in the above mentioned way, links are included in order to form a preliminary concept map.
- The links are labeled with linking words describing the relationships that exist between the concepts.
- The last step is to review the developed concept map and – if necessary – to do changes in structure and/or content.

These steps describe the activities that typically take place in a successful effort of concept mapping. Of course, in practice the actual process will rarely proceed in this clear and sequential fashion (Coffey et al., 2003).

When analyzing and organizing a knowledge domain by means of concept mapping a high degree of freedom is provided. In general, this method does not require following certain required structural properties of the content, such as a tree structure. The visual properties of a graphically represented concept map may or may not emphasize the contents' structural property, as e.g. a hierarchical content can be visually depicted in a variety of ways. Novak (2001) demands that in a concept map the concepts should be represented in a hierarchical fashion, depicting the most general concepts on top of the map and the more specific ones below. However, Novak also indicates that so-called 'cross-links' are an important characteristic of concept maps. Cross-links are relationships between concepts in different regions or subtopics within a concept map. Therefore, the structure of a concept map cannot be strictly hierarchical, but rather semi-hierarchical (Coffey et al., 2003). In general, the requirement of a hierarchical structure has been questioned (e.g. Abrams, 2000; Ruiz-Primo & Shavelson, 1996). At large, the suitability of a certain style will highly depend on the structure of the knowledge domain that is to be presented. There is no need of imposing a hierarchical structure to a concept map if the structure of the content in question is not hierarchical. Therefore, different map structures seem to be appropriate to depict different types of content structures (Ruiz-Primo, Shavelson, & Schultz, 1997). Concept mapping should provide the freedom of choosing the best fitting representation for the respective knowledge domain, without imposing any constraints regarding its structure. A hierarchical concept map can therefore be considered as a special case of a concept map.

Computer-Based Concept Mapping

Today, numerous software tools are available that allow generating concept maps on the computer (see e.g. Coffey et al., 2003; Lanzing, 2004). These tools provide a variety of advantages compared to the traditional paper-pencil construction of concept maps (Lanzing, 2004; Plotnick, 2001). For computer-based concept mapping, it is most beneficial and convenient to use software tools that are especially designed for the purpose of constructing concept maps (Anderson-Inman & Zeitz, 1993).

Concept Mapping Software

By using concept mapping software, concept maps can conveniently be drawn, adapted, corrected, and manipulated. Images and pictures and other resources can be included in concept maps and the space for drawing the map is not limited to a certain size. These software tools allow the easy restructuring of concept maps through such features as drag and drop. Many of them also provide functions for automatically structuring or organizing a generated concept map by choosing different layout styles (hierarchical structure, web structure etc.). One may annotate concepts (and links) and carry out a spell check. Of course, once a concept map has been generated, it can be conveniently stored in digital format, which caters for easy retrieval and reuse. It is possible to export created maps to different electronic formats. This increases the possibilities of using concept maps as a communication instrument. Some software tools provide the possibility of automatically analyzing or scoring a concept map according to different map characteristics (see also Section 'Concept Mapping for Knowledge Assessment'). The following subsections introduce two illustrative examples of common concept mapping software tools.

Inspiration

One of the most popular concept mapping tools that is currently available is Inspiration (www.inspiration.com). The tool is available in Windows and Macintosh versions, and in a handheld version. There is also a version especially for children (K-5), called Kidspiration. The main functions and features of Inspiration refer to organizing concepts, brainstorming, and visualizing ideas, planning and organizing projects. The software offers comprehensive graphical capabilities, providing lots of different shapes, symbols and images. It allows users to easily add or connect other resources, such as Web pages or data files. A special option allows to view a hierarchical outline representation of the concept map. Constructed concept maps can be transferred to a word processor or exported to several image and hypertext formats. All in all, the software constitutes a useful program for generating appealing concept maps.

IHMC CMap Tools

CMap Tools (http://cmap.ihmc.us/) is a software environment that has been developed by the Institute for Human and Machine Cognition (IHMC) (Cañas et al., 2004). With this software concept maps can be constructed individually or collaboratively. Generated maps can be published and shared with others. They can be linked to related concept maps or to other types of media, such as images, videos and Web pages. It is possible to search the Web for information relevant for a concept map. Collaboration is enabled through synchronous and asynchronous communication features. The software is free for use by anybody. It allows for exporting a concept map to an image file, an outline format, a Web page and even an XML file. A concept map may be converted to or imported from a proposition list in text file format. CMap Tools also supports the import of concept maps stored in other formats (outlines, XML etc.).

Other Suitable Software Tools

Software tools that are especially designed for mind mapping, such as MindManager (http://www.mindjet.com) or MindMapper (http://www.mindmapper.com) only support hierarchical tree structures, without any cross-links between different regions of the map. As mind maps feature only associative, unlabeled relations, mind mapping software typically does not provide the opportunity to label the relations between concepts and to make explicit the direction of a relation by arrowheads. These tools are powerful for creating mind maps as special cases of concept maps, but seem less suitable for generating concept maps in general.

General diagramming tools, such as Visio (http://www.visio.com) or Smart Draw (http://www.smartdraw.com), provide the functionality for drawing and labeling boxes and generating links between them. These features can be utilized for generating concept maps. However, not all of these software tools promote or even allow link labeling. Common drawing tools (e.g. CorelDraw) can also be applied for constructing concept maps. This kind of tools provides optimal control over how the concept map should look like and there are virtually unlimited possibilities of creating and displaying their elements. They, however, do not provide the functionality of automatically adapting existing links, when nodes are moved, i.e. the links have to be adjusted manually (Lanzing, 2004). Paint tools are a further alternative that can be used to create concept maps on the computer. These software tools offer almost full creative possibilities and using them is nearly like drawing a concept map with paper and pencil. This fact reveals also the drawback of such programs. Editing, e.g. rearranging, created maps is difficult, as it is not possible to move single elements of the map structure.

Empirical Research on Computer-Based Concept Mapping

Lin, Strickland, Ray, and Denner (2004) compared computer-based concept mapping and paper-based concept mapping as a preparation strategy for a persuasive writing task. It could be shown, that computer-based concept mapping enhances idea generation as well as the total quality of students' concept maps. The performance on persuasive writing, however, was better for paper-and-pencil concept mappers than for students that had generated their maps on the computer – a result that the authors attribute to the applied assessment rubric for the essays. Sturm and Rankin-Erickson (2002) also examined paper-based vs. computer-based concept mapping as a strategy to support essay writing of students with learning disabilities. Students' attitudes toward writing were identified to be significantly more positive for computer-based concept mapping, compared to paper-based concept mapping.

Anderson-Inman and Zeitz (1993) argue that in educational practice students find the process of constructing and modifying paper-based concept maps very difficult. They often resist making changes in a concept map, as it requires much effort – most probably recopying and regenerating the map. Conversely, the application of computers for concept map generation allows for easily modifying and revising concept maps and thus, makes the mapping process more accessible to learners. Empirical evidence argues that computer-based concept mapping is able to overcome frustration and confusion felt by learners when they are expected to generate concept maps and encourages learners to revise their maps (Anderson-Inman & Zeitz, 1993).

Concept Mapping for Instructional Planning

Curriculum development and instructional planning is of great interest at all levels of education, because it involves defining learning goals, scope and sequence of instructional content, learners' interaction with content, and methods for assessing learning outcomes. Instruments applied in this context should effectively support knowledge elicitation and organization, from a range of sources such as textbooks, reports, experts, lecture summaries etc. Curriculum planning instruments should also provide support in analyzing the relationships between curricular elements (Allen, Hoffmann, Kompella, & Sticht, 1993). Concept mapping facilitates the representation of educational domains and can be applied for planning a curriculum or instruction on a particular topic (Coffey et al., 2003; Novak, 1998, 2001). In general, it may aid in developing and organizing a whole course, but also in preparing specific lessons (Zimmaro & Cawley, 1998).

With concept mapping, the structure of a knowledge domain can be made explicit, by using it as a tool for content and curriculum analysis. Concept maps generated in this context may contain both, content-based relations and instructional relations. Content-based relations describe semantic relationships among the concepts to be taught. Instructional relations are relations that provide information on the instructional sequence with respect to these concepts, by uncovering prerequisite relations among them. When utilizing concept maps in the initial curricular planning process, curriculum developer and educators gain a comprehensive understanding of what students need to learn (McDaniel, Roth, Miller, 2005). This may lead to identify areas or subtopics as being trivial, so that they possibly can be dropped from a course, or as being worth to be emphasized during instruction. Concepts can be determined, that are fundamental for more than one topic or knowledge domain. Concept maps help educators to increase their potential of seeing multiple ways of constructing meaning. They help to explain why a particular concept is worth knowing and how it is related to other concepts within and beyond the topic (Allen et al., 1993). Concept mapping used as a planning device for instruction may provide suggestions for an appropriate sequencing of the instructional material (Clark & James, 2004; Martin, 1994; Novak, 2001). The educator is supported in introducing and teaching concepts in an order that allows learners to better incorporate and integrate them with their existing knowledge (Clark & James, 2004). Concept mapping can therefore help to design units of study that are relevant, meaningful, pedagogically sound, and interesting to learners (Martin, 1994). As a result, using concept maps for educational planning can help to improve curricular quality and clarity of teaching.

Instructional Planning Strategies

Allen et al. (1993) defined a six-step model for developing a so-called 'integrated curriculum knowledge map' (see Figure 2). First, concepts or skills that are relevant for the knowledge domain in question are collected from different sources (textbooks, experts, existing course syllabi). These concepts are denoted as 'descriptors of knowledge' and represent the nodes of the map. In a second step, hierarchical relationships between those knowledge descriptors are identified and mapped. After that, heterarchical relationships (e.g. procedural or causal) are identified and created. By considering the concepts and branches in the map it is possible to get an idea of the particularly important concepts of the domain, as those will have many direct links to other concepts. In the fourth step, course development nodes are constructed. For this, the concepts that are relevant for a particular course are identified. Based on these concepts subsumption nodes - involving several more specific nodes - can be built and denoted by a general descriptor. Subsequently, appropriate learning activities and learning resources (e.g. textbooks, demonstrations etc.) are assigned to the various course concepts or topics. Finally, course development nodes can be selected, compiled and sequenced.

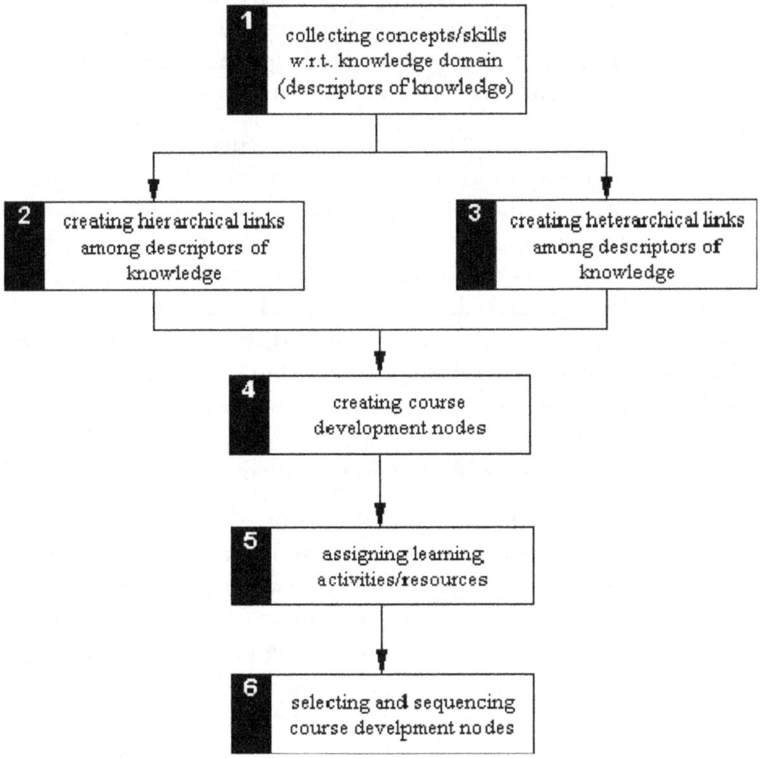

Figure 2: Six-step model for the use of concept mapping in the scope of curriculum development (adapted from Allen, Hoffmann, Kompella, & Sticht, 1993)

Anderson-Inman and Ditson (1999) describe a curriculum-planning strategy that is comprised of three major steps. First, the major concepts of the course in question are identified and represented in a concept map. These concepts constitute single units or lessons of the course. After that, these major concepts are extended by adding key propositions and important examples. When this is completed for the whole course, the educator can elaborate the concepts, propositions, and examples with instructionally appropriate and powerful information. This can be done e.g. by annotating concepts with questions that may be asked during a discussion. Finally, each major concept can be expanded into a more detailed concept map, which can easily be realized with common concept mapping software - by creating and linking submaps or nested nodes, respectively. Such more detailed concept maps are useful for recording planned activities within a lesson. In this way, concept mapping serves for working out the subject matter content and how that content will be translated into lessons.

Clark and James (2004) describe how concept mapping can be applied for planning and organizing a course. Based on lecture summaries a set of concept maps is generated from which an appropriate sequence for teaching the different course topics can be gained. This sequence ensures that new concepts can be linked to already existing knowledge or concepts that have already been presented. By applying this method, the amount of details presented is reduced, thus avoiding a concept overload by concentrating on the essential and most relevant concepts. Proceeding in this way, the resulting course is assumed to allow learners to better retain the knowledge acquired, to exit the course adequately competent, and with a solid foundation for further learning.

Concept maps that are developed in the scope of instructional planning offer a basis for communicating with students (Anderson-Inman & Ditson, 1999). Concept maps depicting instructional content help to make an instruction conceptually transparent to learners. To this end, Novak (2001) suggests to create a global concept map, showing the basic concepts or ideas that are to be taught in a course, and a range of more detailed concept maps representing specific parts or topics of the instruction. This leads to a further application area of concept maps, which is their use as a teaching strategy.

Concept Mapping as a Teaching Strategy

In teaching practice, relying on textbooks poses the general problem of providing only one path through a given domain of knowledge. Novak (2003) emphasizes that the respective linear instructional sequence of learning content makes sense to the book author but not necessarily to learners, i.e. it can never be an optimal sequence for all learners in any class or group. Concept maps can be used as communication tools for presenting information to learners (e.g. Lanzing, 2004) – as an alternative or rather as a supplement to traditional learning material. Concept maps allow for a multitude of paths for reading off the information that is represented and thus provide the learner with the possibility of choosing the most appropriate sequence for him or her. By visualizing the learning content, concept maps help students to see and understand the connections that exist between concepts of a topic. Concept maps may be implemented in the teaching process in several ways. They may be applied as advance organizers in the sense of Ausubel (1968), giving an overview of the content to be learned (e.g. Coffey & Cañas, 2003; Novak, 1998). Another alternative is to present the whole learning content in form of concept maps (e.g. Bernd, Hippchen, Jüngst, & Strittmatter, 2000; O'Donnell, Dansereau, & Hall, 2002). A concept map may also be used as a post organizer providing a summary of the learning material after a learning episode (Hall & Strangman, 2002).

Concept maps are and were often applied for presenting learning material in the context of science education (e.g. Heinze-Fry, 2004; Novak, 2003, 2004). They have a large potential of effectively and appropriately communicating mathematical knowledge (Baroody & Bartels, 2000; Brinkmann, 2003, 2005). Concept maps representing learning material have also been successfully applied in the context of foreign language education (Bahr & Dansereau, 2001, 2005).

It could be shown that concept maps are suitable instruments for reducing communication barriers between persons of different ethnic background. For instance, Dansereau, Joe, Dees, and Simpson (1996) applied concept maps in the context of drug abuse counseling, for presenting client issues to individuals of different ethnicity (African American, Mexican American, White). When counseling was supplemented by the use of concept maps, it was more effective compared to traditional counseling. Therefore, the utilization of concept maps seems to help in reducing "cultural [and] racial ... communication barriers by providing a visual supplement and a common language that enhances ... interchanges [between individuals]" (Dansereau et al., 1996, p. 363).

The following subsections indicate different uses of concept maps as a teaching strategy and related empirical research.

Concept Maps as Advance Organizers

An advance organizer provides a global overview of the material to be learned and can take a variety of representation forms like text passages or summaries, text illustrated by pictures, graphical or map representations (Coffey & Cañas, 2003). Advance organizers were propagated by Ausubel (1968), who claimed that their use fosters meaningful learning. They direct learners to basic concepts that they have possibly already available in their knowledge. Advance organizers help learners to relate new knowledge to existing knowledge and thus provide an anchor for learning new content (Novak, 1998). In this way advance organizers help learners to bridge the gap between already existing and to be learned knowledge.

Concept maps can be utilized as powerful advance organizers in teaching, making explicit the general, superordinate concepts of the topic to be learned and how they are interrelated (Novak, 1998). Heinze-Fry (2004), for example, used concept maps as advance organizers for lectures in introductory science courses. She presented a general overview map for a whole course and more specialized ones for introducing each new chapter.

Willerman and Mac Harg (1991) empirically investigated concept maps as advance organizers in middle school physics. They assumed that presenting them at the beginning of a unit would improve performance in a post test. The experimental group was presented with the concept map, accompanied by explanations of the relationships among the concepts. These learners were asked to construct concept maps by copying the teacher's example. The control group only received an introduction that stated the unit's objectives and included interesting questions in order to foster motivation. Results indicated that the use of an advance organizer concept map led to a significantly better performance in a test at the end of the unit. The map gave the learners a helpful overview of the key concepts and their relationships before they were exposed to more in-depth instruction. Thus, this study provides empirical support for making use of concept maps as advance organizers.

Concept Maps Representing Learning Content

Concept maps may not only be used as an overview of subsequent learning material, but they may also serve for presenting the learning content of a whole unit or course – in face-to-face as well as in distance education. Due to their visual representation, concept maps are assumed to effectively support learning and memorizing. They are able to represent complex and extensive information more economically and understandable than texts do. Concept maps make the macrostructure of a knowledge domain more salient (O'Donnell et al., 2002) and can be beneficially applied for communicating complex ideas and abstract concepts in e-learning as well as in classroom teaching (e.g. Dabbagh,

2001; Plotnick, 2001). In this way, they facilitate gaining a holistic and networked understanding of a knowledge domain. Concept maps that represent learning material are typically maps created by experts. The underlying conception of teaching is that students have to acquire this representation (Bruillard & Baron, 2000). Novak (1990) advises against presenting learners with prepared concept maps with only the request to learn the maps and their content by heart and to give an analogue reproduction. Rather, a provided concept map should be used in a more elaborated and conscious way for learning. O'Donnell et al. (2002) state, that although learning from expert concept maps seems to be "suggestive of a focus on reception learning, the learner must construct the relationships that link two or more nodes" (p. 75). To ensure a positive learning effect Bernd et al. (2000) point out the importance of working through a concept map in a systematic and elaborated way – in correspondence to Aebli's (1985) 'working through' as one of twelve fundamental forms of teaching.

In an empirical investigation, Hall and O'Donnell (1996) presented their students with either concept maps or traditional texts. The results showed that those learners who had studied from the maps scored significantly better in a free recall test than those who had studied from the texts.

Bahr and Dansereau (2001) emphasized the potential of concept maps to communicate information in two or more languages simultaneously, because of their almost universal grammatical structure. A bilingual (or multilingual) concept map can be generated by including more than one language representation of a concept in a concept node. The researchers used such bi-lingual maps for presenting German-English vocabulary and found that this presentation format led to better recall compared to lists of word-pairs.

It was shown, that the presentation of learning material through concept maps (or similar representations) alternatively to texts is especially beneficial for learners with low verbal ability or skills (for an overview see e.g. O'Donnell et al., 2002). For example, Patterson, Dansereau, and Wiegmann (1993) compared high and low verbal ability students that were taught by either concept maps or texts. They found that low verbal ability learners performed better on recall measures when they were provided with a map representation.

Concept Maps as Post Organizers

Using concept maps as post organizers means to present them after a learning episode, as an integrative tool. In this case, the concept map summarizes and consolidates what has been taught. Advance and post organizers may be combined as in Walker, King, and Cordray (2003). Their lectures began with presenting the advance organizer which contained the major concepts of the chapter and was concluded with providing a more detailed concept map summarizing the covered topics.

Empirical evidence for the effectiveness of concept map post organizers has been provided by Jüngst (1995), who conducted an investigation in the scope of an introductory university course in pedagogy. After presenting learning content in a traditional lecture, students received and worked through either a concept map or an analogous text as a summary. The results in ensuing knowledge tests favored the concept maps group.

Further Uses of Concept Maps for Teaching

Another alternative for educators to apply concept mapping as a teaching strategy is to create a concept map interactively with learners. This means, for example, that the teacher poses a question or raises a discussion and captures the students' responses and reactions in form of a concept map (Dabbagh, 2001) – in contrast to concept mapping as a learning strategy, where learners construct concept maps themselves (see Section 'Concept Mapping as a Learning Strategy'). Guastello, Beasley & Sinatra (2000) applied concept maps in a similar way in teaching low achieving students. Concept mapping was realized as an instructional strategy in addition to a traditional read-and-discuss instructional technique. All students read a textbook chapter section by section together with the teacher and discussed the meaning of the text after each passage. One group of learners constructed a concept map with the teacher's assistance, based on the discussion after each section. In a post-test assessing text comprehension, the learners in the concept mapping group gained significantly better scores compared to the pure read-and-discuss instruction.

A concept map prepared by the educator representing the key ideas and their relationships of a particular topic can also be used as a kind of 'conceptual scaffolding' for learners (Novak, 2003). This means, the provided concept map may serve the learners as a starting point for constructing their own, more elaborated concept maps during the learning episode (for further details see Section 'Concept Mapping as a Learning Strategy').

Concept Mapping as a Learning Strategy

Many researchers favor the use of concept mapping as an active learning strategy, emphasizing that those who construct a map get most out of it (e.g. Anderson-Inman & Zeitz, 1993; Novak, 1990, 2003; Lumer & Hesse, 2004). Constructing concept maps is an active process that requires to get intensively engaged in a topic, and to elaborate and reflect information. The activity of concept mapping helps learners to become aware of the process of linking new knowledge elements to already existing ones and to consciously promote this process. Therefore, concept mapping constitutes a task that actively engages learners in searching connections between their existing understanding and new knowledge and is able to encourage meaningful learning and a deeper understanding of a subject matter (Cardellini, 2004; Novak, 2003). During the active process of generating concept maps, learners can detect their own knowledge as well as gap's in their knowledge. Concept mapping poses an exploratory activity fostering explication and reflection and helps to develop auto-monitoring techniques and critical thinking (Dabbagh, 2001). The learner becomes "more aware of the necessary regulation of her learning processes in relation to the abilities to acquire" (Bruillard & Baron, 2000, p. 333).

The volume of research on concept mapping as a learning strategy is immense. Concept mapping has been applied in virtually every field of study, including natural sciences (e.g. Chang, Sung, & Chen, 2002; Heinze-Fry, 2004; Lumer & Hesse, 2004), mathematics (e.g. Afamasaga-Fuata'i, 2004; Brinkmann, 2003, 2005; Bolte, 1999; McGowen & Tall, 1999), psychology (e.g. Jacobs-Lawson & Hershey, 2002), foreign languages (e.g. Bahr & Dansereau, 2004), patient education (e.g. Marchand, D'Ivernois, Assal, Slama, & Hivon, 2002), and history (e.g. Karasavvidis, 2004).

Empirical Research on Concept Mapping as a Learning Strategy

The supportive effect of concept mapping during the learning process has been investigated intensively. A majority of studies addressed the objective learning outcome of concept mapping, but there are also studies that considered the affective outcome, i.e. the subjective reactions of learners. Concerning the objective learning outcome, the effect of concept mapping on retention or recall of learned information, but also the transfer to problem solving has been examined.

Studies often compared the use of concept mapping during learning with traditional instructional methods. Czerniak and Haney (1998), for example, compared traditional instruction in a physical science course with instruction complemented by concept mapping. The effect on achievement and anxiety toward physical science was investigated. It could be shown, that concept mapping increased achievement in a post-test and decreased anxiety compared to learners that had not used that learning strategy. Similar results in favor of concept mapping were reported by Jegede, Alaiyemola, and Okebukola (1990), who compared the addition of concept mapping to instruction with traditional instruction in biology. Nicoll, Francisco, and Nakhleh (2001) addressed the question whether the use of concept mapping would help learners to produce a more interconnected knowledge compared to traditional instruction in chemistry. In order to determine the degree of interconnectedness of learners' knowledge a structured interview was carried out after instruction. The interview aimed at eliciting concepts of the subject matter and relationships between them. Those students who had used concept mapping knew more concepts as well as relationships and had no more erroneous relationships than the students from ordinary instruction. Compared to traditional note taking a mapping strategy also leads to higher scores in retention-tests (Reader & Hammond, 1994) and enhances recall of high and medium importance information (Denner, 1992). A few studies, however, found no difference between the use of concept mapping and other strategies, such as concept defining (Spaulding, 1989, as cited in Coffey et al. 2003) or outlining (Lehman, Carter, & Kahle, 1985).

Another possible strategy of applying concept maps as a supporting tool for learning is to provide students with a partial (or skeleton) concept map, where part of the concepts and relations are provided and the other part is set as blanks. This partial concept map can be used as a scaffold for students for generating their own concept maps – by filling in the blanks (see also Section 'Concept Mapping for Knowledge Assessment'). This method is assumed to reduce cognitive load compared to constructing a concept map from scratch since it provides a referent knowledge structure to learners (Chang, Sung, & Chen, 2001). Chang et al. compared the learning effectiveness of creating concept maps vs. completing skeleton maps in the field of biology. Both methods were based on provided lists of concepts and relations that should be used. A control group of learners generated concept maps with paper and pencil without receiving any feedback on recommended revisions. Learning achievement of those learners working with the skeleton maps was better than when generating concept maps on the computer as well as with paper and pencil.

Concept mapping can also be applied as an instrument for enhancing the process of problem solving (Lanzing, 2004; Plotnick, 2001). In this regard, concept mapping is utilized as a creativity instrument in the problem solving process, for collecting and generating alternative solutions and options. A concept map makes it possible to visualize and clarify ideas and to trigger associations leading to new ideas (Lanzing, 2004).

Concept Mapping as a Collaborative Learning Strategy

Concept mapping as a learning strategy is not restricted to individual use. Rather, concept maps may be jointly generated by two or more learners. In such a setting, concept mapping constitutes an efficient means for facilitating and supporting cooperative and collaborative learning (e.g. De Simone, Schmid, & McEwen, 2001; Mandl & Fischer, 2000; Novak, 2001). Collaborative concept mapping can be used by learners for elaborating and acquiring knowledge on a particular knowledge domain. It may also be applied in problem solving processes in order to clarify and articulate the knowledge that is needed for solving the problems in question, and for finding a solution (Novak, 2001). In this context, a concept map acts as a kind of 'problem space', providing an appropriate means for structuring and coordinating information (Mandl & Fischer, 2000). The quality of processes in co-constructing meanings and solutions as well as the outcomes of collaborative learning can be positively affected by the application of a mapping technique (Bruhn, Fischer, Gräsel, & Mandl, 2000).

Learners' discourse as it is initiated by a collaborative concept mapping task has several features (Van Boxtel, Van der Linden, Roelofs, & Erkens, 2002):

- Learners are engaged in articulating and communicating their thoughts. They explain their own conceptions of and experiences with the subject matter and therefore are able to gain a greater conceptual clarity. They also become aware of existing knowledge gaps. Providing concepts for the map construction helps learners to pay attention on the key principles of the knowledge domain.

- Learners are engaged in elaborating their conceptual knowledge. By the necessity of articulating their ideas, learners are enabled to elaborate, question and criticize them. Collaborative concept mapping elicits negotiation, which is characterized by asking and answering questions and resolving disagreements.

- Learners are engaged in co-constructing meanings. Peers working on a collaborative concept mapping task have to create and sustain a common understanding. They need to have a shared meaning of task, concepts, relationships among them, and strategies to use. All parts need to contribute in answering questions, resolving conflicts, and constructing reasons in order to establish a mutual understanding.

Empirical Research on Collaborative Concept Mapping

Esiobu and Soyibo (1995) applied concept mapping together with a second mapping strategy in individual or group situations and compared those learners with a control group that did not use any mapping technique. Learners that used concept mapping had better learning outcomes than the control group, and there was an advantage for group learning, compared to individual concept mapping.

In their research, Van Boxtel et al. (2002) conducted empirical investigations in the knowledge domain of physics, having learners create paper-based concept maps in pairs by using a given set of concepts. The authors concluded, that concept mapping as a collaborative learning activity is effective and leads to significant learning gains. Additionally, learners' interactions were recorded and analyzed. It was found, that learning outcomes seem to be related to the quality of interaction among the learners. The more focused the interaction was on concepts of the subject matter and the more elaborative this discussion was, the better was the learning outcome.

Concept Mapping for Knowledge Assessment

A concept map created by an individual reflects the understanding of this very person in a given domain of knowledge. Therefore, concept mapping can not only be successfully applied for teaching and learning, but it can also be utilized as an instrument for evaluation, i.e. for assessing learners' knowledge (e.g. Novak, 2001; Lanzing, 2004; Ruiz-Primo, 2004). By asking individuals to generate a concept map on a particular knowledge domain, they are engaged in constructing a semantic network that mirrors their understanding of the respective domain (Dabbagh, 2001). Concept mapping for educational assessment (in short, concept map assessment) can serve two purposes, formative or summative evaluation (Coffey et al., 2003; Lanzing, 2004). Formative assessment means asking learners to construct concept maps repeatedly at various points during the instructional and learning process. Assessing the learners' understanding at a certain point in time through the resulting maps can form a basis for evaluating and modifying teaching or teaching material. Summative assessment means to use a concept mapping task at the end of an instructional unit. This refers to the use of concept mapping as achievement test, for assessing performance of learners and assigning grades.

Concept mapping provides a set of procedures that can be used to assess important aspects of a learner's conceptual knowledge and its organization (Ruiz-Primo, 2000). According to Ruiz-Primo and Shavelson (1996) concept map assessment is characterized by three features:

- a concept mapping task, that prompts a learner to communicate his/her knowledge structure of a content area
- a response format, prescribing the type of responding to the task
- a scoring system that is used for evaluating the learner's response, i.e. concept map

A whole range of procedures for posing a concept mapping task is available. Theses different procedures vary considerably with respect to the extent to which they impose constraints or provide information. Ruiz-Primo (2000, 2004) introduced the notion of a dimension of 'directedness' for describing the varying degree of constraints that a concept mapping task entails. A low-directed task does not put any constraints but leaves the learner complete freedom in deciding which and how many concepts are used for the concept map, how they are related and how these relationships can be labeled. A higher level of directedness characterizes a map creation task that e.g. provides an open-ended list of concepts that is to be used for generating the map. A very high-directed task provides a learner with pre-defined concepts and relations and the map structure including connecting lines (i.e. map completion), and thus is highly restrictive.

The following sections provide an overview of concept mapping tasks and techniques. For illustrating the described techniques, example tasks are presented which are based on the concept map describing what a concept map is (presented in Figure 1). Please note that generally all these procedures are applicable in a paper-based form, but can be or have already been realized as computer-based procedures. For example, there exist already software tools that do not only allow carrying out concept mapping tasks for knowledge assessment, but also provide functionalities for a computerized analysis of the gathered concept maps, e.g. calculating structural measures of self-constructed concept maps or scoring the achievement on filled in skeleton-maps (e.g. Aidman & Egan, 1998; Bernd et al., 2000; Chang et al., 2001).

Map Creation

The 'construct-a-map' or 'map creation' technique requires generating a concept map concerning a particular knowledge domain from scratch (Ruiz-Primo, 2000; Schau & Mattern, 1997). In practice, this method varies in its degree of directedness according to how much information is provided. The concepts and/or relations to be used for generating the concept map may have to be generated or, alternatively, may be provided. If concepts and relations are allocated, this list may either be open-ended or fixed. Moreover, information regarding the required structure of the concept map (hierarchical or non-hierarchical) may be provided or not. Table 2 provides an example of a concept mapping task, where a concept list is provided. A map creation task may involve drawing the concept map for instance by hand with paper and pencil, on the computer, or by arranging note cards. Of course, such a concept mapping task may address an individual learner, but also groups of learners.

In order to assess learners on the basis of their constructed concept maps and to make them comparable to each other, they have to be evaluated through a specific scoring system. Until today, numerous scoring systems have been proposed (e.g. McClure, Sonak, & Suen, 1999; Novak & Gowin, 1984; Ruiz-Primo et al., 1997; Rice, Ryan, & Samson, 1998; Rye & Rubba, 2002).

Table 2: Example of a map creation task
Have a look at the concepts listed below. They focus on the knowledge domain of concept maps. Create a concept map describing what a concept map is by using the terms provided. You may arrange the concepts in any way you want. For relating concepts, use arrows (indicating the direction of the relation) as well as labels (describing the kind of relationship).
• arcs • concepts • concept map • graph • linking words • nodes

In general, three different scoring strategies can be distinguished for evaluating a constructed map (Ruiz-Primo & Shavelson, 1996). One possibility is to score the components of a generated concept map. This means to evaluate a concept map according to different map characteristics, such as number of correct propositions, proposition accuracy, crosslinks, hierarchy levels, examples etc. The scoring system proposed by Novak and Gowin (1984), for instance, counts the number of valid propositions, levels of hierarchy, branchings, cross-links, and examples. Each of these counts is given a weight, and the sum of points assigned in this manner forms the score for the concept map. A second and often applied strategy for evaluating learner-created concept maps is based on the comparison with an expert map (also called master map). This means, the overlap between the learner and the expert map is scored, most often by calculating a kind of similarity-index (e.g. Chang, Sung, Chang, & Lin, 2005; Marshall, Chen, & Madhusudan, 2006; Ruiz-Primo, 2000; Takeya, Sasaki, Nagaoka, & Yonezawa, 2004). Ruiz-Primo (2000), for example, used a so-called convergence score, which is given by the proportion of accurate propositions in the learner concept map out of the total possible valid propositions in the expert map. A third strategy is to use a combination of both component scoring as well as expert map comparison. In general, different scoring systems emphasize different aspects of a concept map, e.g. the elaborateness of a map or the validity of the represented knowledge (Stoddart, Abrams, Gasper, & Canaday, 2000). Therefore, different scoring systems probably lead to different results when applied and may even refer to different abilities. To sum up, the literature offers a variety of different scoring systems. The problem is, however, that until now none of them is universally accepted (Schau, Mattern, Zeilik, Teague, & Weber, 2001). In fact, the possible scoring strategies still require further research in order to estimate their ability of providing useful and valid information about a person's knowledge in a particular field (Ruiz-Primo & Shavelson, 1996).

Learners that are going to participate in a map creation task for knowledge assessment first need to learn how to construct concept maps. In other words, they should receive some training (e.g. Hibberd, Jones, & Morris, 2002). This is a time-consuming process and might result in frustration and rejection of the method itself. The quality of concept maps also highly depends on the learners' individual communication and verbal skills (Schau et al., 2001). In general, this method leaves the individual the highest degree of freedom to construct a concept map in the way that he or she wants. Concept maps arising from tasks of this type may therefore most comprehensibly be seen as a graphical representation of an individual's personal understanding.

Map Completion

The 'fill-in-the-map' or 'map completion' method constitutes another possibility of posing a concept mapping task to individuals (Ruiz-Primo, 2000; Schau et al., 2001). In this case, a concept map is provided, where all or some of the concepts and/or relations have been left out (see Figure 3 for an example). The blanks in this partial map (or skeleton map) have to be filled in, either by generating the respective words or by selecting them from a list, which may include distractors. Although also other scoring schemes are possible, most commonly the responses are simply scored as correct or incorrect.

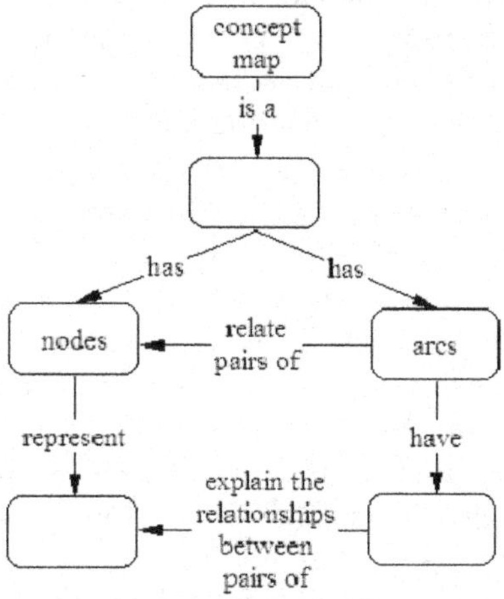

Figure 3: Example of a skeleton map (adapted from Ruiz-Primo, 2000)

Some researchers claim that this method should be preferred to the map creation technique, as the latter is assumed to impose a too high cognitive demand (Schau, et al. 2001). Map completion tasks can easily and quickly be administered. However, this method poses the problem that only a part of the knowledge can be queried, as it is necessary to provide at least a small part of the concepts and/or relations in the map that has to be filled in. As a representation of the structure of the knowledge domain is provided, a person does not have to create an individual structure representing his/her personal understanding of the domain. To sum up, map completion tasks incorporate a high degree of constraints and provide the learner with a lot of information, as the structure of the concept map and part of its content is allocated. The specific advantage of this method lies in its convenience to administer and score.

Relatedness Ratings

One further technique that can be classified as a concept mapping task makes use of relatedness ratings. This method is characterized by a two-stage indirect approach (Schau & Mattern, 1997). Individuals are asked to rate the degree of relatedness between pairs of previously defined concepts on a numerical scale (see Table 3 for an example). Proceeding in this way, only the existence and strength of a relationship between two concepts is queried, but not the nature or label of this relation. Each item in such a task, i.e. each pair of concepts that has to be rated, can be considered as an isolated proposition with an unlabeled link (Schau et al., 2001). The relatedness ratings of an individual can be represented visually in form of a concept map. This is done through applying a mathematical algorithm, e.g. by using the Pathfinder software (Schvaneveldt, 1990), which allows to create a network representation based on the relatedness matrix. Such maps that are based on relatedness ratings do not include labeled relations, but only indicate the degree of relatedness through the distance between two given concepts. The resulting maps are therefore no concept maps in a general sense. An extension of the original method, though, is able to overcome the drawback of unlabeled links (Shavelson, Ruiz-Primo, & Wiley, 2005). To this end the network representation resulting from the relatedness ratings can be provided to the individual by asking him/her to label the relations between the concepts and, if necessary, to add or remove relations. The resulting maps can actually be considered as concept maps.

Table 3: Example of a relatedness rating task

Judge the relatedness of each pair of concepts by checking one of the provided numbers! ⓪ indicates that the concepts are unrelated, ⑥ indicates that the concepts are closely related.

Concept Map	Graph	⓪ ① ② ③ ④ ⑤ ⑥
Concept	Proposition	⓪ ① ② ③ ④ ⑤ ⑥
Arc	Linking Word	⓪ ① ② ③ ④ ⑤ ⑥
Concept	Linking Word	⓪ ① ② ③ ④ ⑤ ⑥

Applying the original technique of relatedness ratings an individual is not supported by a map representation, as it is the case with the map creation or map completion method. When judging the relatedness between a pair of concepts a learner possibly is not aware of the context or area of knowledge, in which the respective concepts are embedded (Schau et al., 2001). Implementing an additional step of querying relation labels as indicated above, however, entails the support of a visualized network representation. All in all, the technique of relatedness ratings imposes a quite high level of directedness in the sense of Ruiz-Primo (2000, 2004), as the concepts on which ratings are based are previously defined. In case of adopting the mentioned extension of the method, a higher degree of freedom is realized. Furthermore, as an additional benefit the extended method provides labeled relations.

Proposition Correct-Incorrect Discrimination Task

Yet another kind of concept mapping task utilizes a concept map's representation in form of a list of propositions. The propositions of an expert concept map representing a particular knowledge domain are presented to individuals, requiring judging each proposition in a kind of proposition discrimination task. This method has been implemented in the concept mapping software 'Cmap Pro' (Bernd et al., 2000) for assessing previous knowledge before starting a learning session. In this case, an individual has to judge each proposition, whether it is contained in his or her personal understanding or not. An extension of this method was suggested and applied by Steiner (Steiner, 2004; Albert & Steiner, 2005). Using this extended technique, the propositions of the concept map are combined with distractor items (incorrect statements) and presented in form of a correct-incorrect discrimination task. That is, each statement has to be judged as either correct or incorrect. For reducing the risk of lucky guesses confidence ratings may be included, requiring for each statement to specify the degree of confidence that the given judgment is correct (see Table 4 for an example).

Based on the answer pattern of an individual, a graph representation of a concept map can be established, representing the personal understanding of this person. This concept map contains all those propositions that have been judged by the person as being 'correct'. Notice that, of course, limited by the presented distractor items, such an individual concept map may also include misconceptions in case that distractor items have been erroneously judged as 'correct'. Based on such a concept map representing individual understanding, an additional phase of answer checking can be realized. This means, given the concept map of a person, he or she could be provided with the opportunity to check the answers and make corrections, if necessary. Scoring can be done by simply counting the number of correctly judged propositions, but other scoring procedures apply, too. Collected confidence ratings can be utilized for additionally weighting the respective scores.

Of course, this technique features also a high level of constraints, as the content of the expert concept map – together with distractor content – is provided. In contrast to the map completion method, this technique allows for querying the whole knowledge contained in the expert map. As there is an omnipresent risk of guessing, the knowledge level of a person might be overestimated. Therefore, confidence ratings should definitely be included, allowing for a signal detection theoretical analysis of answer patterns. The answer pattern received from the confidence-rating method allows for creating a ROC curve for each individual, which can be used to derive a sophisticated performance measure (see e.g. Swets, 1996; Stanislaw & Todorov, 1999). In sum, this type of concept mapping task seems to be convenient and easy to use, and does not need any prior training.

Table 4: Example of a proposition correct-incorrect discrimination task including confidence ratings

A concept map is a graph.	☐ correct	☐ +	☐ ++	☐ +++
	☐ incorrect	☐ –	☐ – –	☐ – – –
Nodes represent relationships.	☐ correct	☐ +	☐ ++	☐ +++
	☐ incorrect	☐ –	☐ – –	☐ – – –
Arcs have linking words.	☐ correct	☐ +	☐ ++	☐ +++
	☐ incorrect	☐ –	☐ – –	☐ – – –
Nodes relate pairs of concepts.	☐ correct	☐ +	☐ ++	☐ +++
	☐ incorrect	☐ –	☐ – –	☐ – – –
Arcs relate pairs of nodes.	☐ correct	☐ +	☐ ++	☐ +++
	☐ incorrect	☐ –	☐ – –	☐ – – –

Comparing and Choosing Concept Mapping Tasks

Although there are various possibilities for posing a concept mapping task the most frequently applied method is map creation. As shown in the preceding sections, the different types of concept mapping tasks impose different levels of constraints and provide different amount of information to individuals. Also within one type of mapping task the extent of allocated information and constraints may vary. Therefore, several empirical studies tried to compare different assessment methods, addressing the questions whether they measure the same or different abilities, whether they lead to different levels of performance or which method is best suited for knowledge assessment. Comparing the map creation and the map completion technique it was shown that these two mapping techniques impose different cognitive demands on students and may provide different pictures about students' knowledge (Ruiz-Primo, 2000; Ruiz-Primo, Schultz, Li, & Shavelson, 2001; Ruiz-Primo, Shavelson, Li, & Schultz, 2001; Shavelson et al, 2005). Based on their research, Ruiz-Primo et al. argue that the map creation technique is the 'gold standard' of concept map assessment. They state, that map creation is able to more accurately reflect differences of learners' knowledge, to capture learners' partial understanding as well as misconceptions, and to elicit more higher-order cognitive processes. Yin, Vanides, Ruiz-Primo, Ayala, & Shavelson (2005) compared two variations of map creation using a set of provided concepts, either by creating linking phrases or by selecting linking phrases from an allocated list. It was found that the two techniques cannot be considered equivalent. The mapping task requiring from learners to choose linking phrases on their own, appeared to be better for capturing learner knowledge. On the other hand, scoring can be done more efficiently when using a list of pre-defined linking phrases. Thus, this technique seems to be better suited for assessments on a large scale.

In general, the selection of a particular type of concept mapping task to be applied for knowledge assessment will highly depend on contextual and situational factors, as for example time demands, available resources, scope of the knowledge assessment (assessing knowledge of one, several, or many individuals) etc. In case, that, for instance, knowledge assessment is to be carried out on a large scale with a great number of learners, a method that is easy to administer and analyze, such as the correct-incorrect discrimination task, will be most appropriate. The assessment objective will also heavily influence the choice of a particular form of concept mapping used for assessment. For example, if comprehensive information on the knowledge and its structural organization of learners is to be elicited, the map creation method seems most appropriate. If on the contrary the aim is to explore learners understanding of the relationships that exist between specific concepts of a knowledge domain, the map completion method (with blanked relations) or the extended procedure using relatedness rating appear suitable. Thus, it seems clear that differing assessment objectives and context factors will influence the choice of the particular type of assessment.

Reliability and Validity of Concept Map Assessment

Until today, numerous studies have been conducted that aimed at providing reliability and validity information of different concept map assessment techniques (Coffey et al., 2003; Ruiz-Primo & Shavelson, 1996). These studies basically addressed map creation and map completion. Regarding reliability, studies predominantly investigated inter-rater reliability. It was shown, that student generated concept maps can be consistently scored by different raters, even when complex judgments are required (Ruiz-Primo, 2000; Shavelson et al., 2005). Moreover, different scoring systems for evaluating generated concept maps were compared with respect to their reliability (McClure et al., 1999, Ruiz-Primo, 2000). Results showed that the selection of the scoring method is likely to have an effect on the score reliability. Regarding validity of concept map assessment it was demonstrated that concept map scores change over the course of instruction as it would be expected (Martin, Mintzes, & Clavijo, 2000; Pearsall, Skiper, & Mintzes, 1997). It was shown that concept map assessment actually is able to distinguish between groups of people that it is expected to distinguish. This means, persons with known differences in their degree of knowledge can be differentiated by concept map assessment (Schau et al., 2001; West, Pomeroy, Park, Gerstenberger, & Sandoval, 2000). In order to give evidence of the construct validity of concept map assessment, concept map scores were correlated with other measures of achievement. In general, moderate to high correlations between concept map scores and multiple choice scores were found (Schau et la., 2001; Rice et al., 1998; Ruiz-Primo, 2000). Furthermore, there is evidence for a strong interrelation between concept map achievement and course grade (Schau et al., 2001). When using standardized tests as a validation criterion, findings are inconsistent. There are studies reporting a strong correlation (Rice et al., 1998) but also studies that did not find any interrelation (West et al., 2000). In sum, the research efforts on reliability and validity issues indicate that concept map based knowledge assessment falls within acceptable ranges from a psychometric point of view (e.g. Coffey et al., 2003; Shavelson & Ruiz-Primo, 2000; West et al., 2000).

Information Gained from Concept Map Assessment

Concept mapping tasks can be effectively applied for knowledge assessment. Concept mapping procedures are able to provide important information about conceptual knowledge and allow to gain information about persons' understanding that cannot be gathered by traditional assessments (Williams, 1998). Therefore, concept mapping is not only suitable for assessing knowledge as a snap-shot of a learner's knowledge, for example after a learning unit or at the end of a course. In fact, concept mapping is also an appropriate tool for monitoring the growth or change of knowledge. This means, concept maps permit observing and recording changes in a person's knowledge over time (Novak, 2003). Concept mapping can for instance be applied as an ongoing evaluation across an instruction. This makes it a useful tool for monitoring the increase of complexity of learners' knowledge structure, as it occurs by integrating new knowledge with existing knowledge (Coffey et al., 2003). A variety of empirical investigations applied concept mapping as an instrument for following changes in knowledge in versatile knowledge domains, such as physics (Novak & Musonda, 1991), chemistry (Liu, 2004), biology (Pearsall et al., 1997), climate change science (Rebich & Gautier, 2005), biomedical engineering (Walker & King, 2003), mathematics (McGowen & Tall, 1999), and diabetic patient education (e.g. Marchand et al., 2002). On the whole, these studies proved the potential of concept mapping for representing changes in knowledge over time.

Concept mapping has also proven to be a suitable method for assessing progress of learning and improvement in performance in complex, dynamic knowledge areas (e.g. engineering design, medical diagnosis), where a variety of solutions and solution approaches exist. In case of such complex problem tasks it is not possible to directly assess a person's knowledge and performance in comparison to a standard solution. A feasible approach is instead to collect information on problem representations and conceptualizations from experts by using concept maps. These maps can then serve as a point of reference for determining the level of understanding of non-expert problem solvers, which are obtained through concept mapping, too (Spector, 2006).

An important by-product of using concept maps for eliciting knowledge is their potential of detecting and uncovering misconceptions that a learner may have (e.g. Lanzing, 2004; Novak, 2002; Plotnick, 2001). Concept maps that are generated by learners express their personal understanding and thus may help the instructor to diagnose misconceptions and incomplete conceptions, which can then be directly fed into remedial actions. Information regarding misconceptions and their clarification usually cannot be provided by traditional assessment methods (Roberts, 1999). Mintzes, Wandersee, and Novak (2001) argue that concept mapping provides the perhaps most powerful method of knowledge assessment that is available "for exploring and documenting the structural complexity and propositional validity of knowledge in scientific domains" (p. 119). They draw attention to the necessity of incorporating concept mapping into the instructional process itself, too, if the potential of this instrument is to be fully realized. This means instead of using concept mapping purely for knowledge assessment, it should also be applied as a learning (and/or teaching) strategy (see Sections 'Concept Mapping as a Learning Strategy' and 'Concept Mapping as a Teaching Strategy').

Concept Mapping and E-Learning

A collection of interlinked Web pages can be described as a network of nodes and links, and thus its structure is very similar to a concept map. This structural similarity makes obvious why it is natural to apply concept maps in this context (Bruillard & Baron, 2000; Lanzing, 2004; Plotnick, 2001). In an educational framework concept mapping may be utilized for designing e-learning, but also for supporting learners to efficiently navigate through learning material. Moreover, concept maps can be used for deriving prerequisite structures among learning objects and competencies, which can form the basis of adaptive e-learning systems.

Concept Mapping and E-Learning Design

In general, designing an e-learning course constitutes an ill-structured problem. Concept mapping can support the phases of problem definition, idea generation, and selection within the design process (Stoyanov, 1997). In the planning process of an e-learning environment, concept mapping can help designers in defining characteristics and requirements. Later in the design process, when building the course, the structure of knowledge that is to be transmitted is a central issue. In order to create an e-learning course that is equipped with a clear structure, authors need to reflect upon the structure of the knowledge domain in question. When organizing the knowledge domain and having available a set of materials that is poorly structured or even unstructured, the application of concept mapping appears very useful, again. This means, when authors cannot rely on existing documents that are well-structured, concept mapping is a valuable instrument for designing the course. Thus, concept maps are able to facilitate the organization of information and knowledge in course planning and design (Bruillard & Baron, 2001; Lanzing, 2004) (see also Section 'Concept Mapping for Instructional Planning').

Stoyanov and Kirschner (2004) report empirical evidence for the usefulness of applying a variant of concept mapping for effective e-learning design. Specialists in the domain of educational technology were asked to produce ideas for developing an adaptive e-learning environment for active learning in a multistep procedure. The mapping approach proved effective for defining the characteristics of the adaptive learning environment, such as, desired functionalities, requirements, and ideas regarding instructional design. A concept mapping method could also be efficiently applied for the next step in the design process, when concepts and ideas have to be transformed into concrete design solutions.

Concept Maps as Navigation Instruments

A well designed navigation instrument is critical for the success of e-learning (Greenwood, Phillimore, & Wang, 2002). Aside from their capability in assisting e-learning design, concept maps may also serve as navigational tools (concept map browsers) in the context of Web based learning. This means, a concept map is included in the e-learning environment, providing an overview and navigational interface for the respective knowledge domain and featuring clickable concept nodes that are linked with the indicated topic (Coffey et al., 2003; Lanzing, 2004; Plotnick, 2001; Shapiro & Niederhauser, 2004). In this way, the concept map provides an advance organizer presenting the key concepts of the domain (see also Section, 'Concept Mapping as a Teaching Strategy'). These concepts are hyperlinked with resources providing learning material on the respective topic (see Figure 4 for an example).

Traditional navigation interfaces in Web based learning are often linear outlines of key concepts or keywords. On the contrary, a concept map constitutes a non-linear display of a navigation interface (Hall, Balestra, & Davis, 2000). As such, a concept map indicates also connections among the key concepts and gives deeper insight in the structural or-

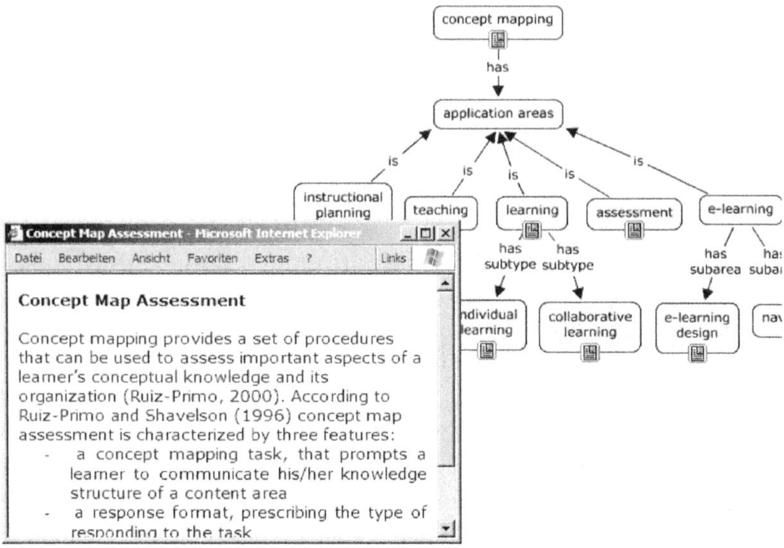

Figure 4: Example of a concept map browser for a learning environment

ganization of a knowledge domain. Therefore, concept maps seem to be a good representational format for an interface displaying hyper-linked pages, and may help learners to find an appropriate way through the learning material. A navigational concept map may foster active processing of the information provided, but at least it provides easier and less frustrating access to information (Coffey et al., 2003).

In the empirical research, the usefulness of concept map browsers compared to other navigation interfaces has been examined. Hall et al. (2000) applied a traditional outline interface vs. a concept map browser in an online learning environment. Learners' perception of the pedagogical effectiveness of the interface differed depending on the provided navigation interface and was much more positive for the concept map condition. However, there was no discriminative effect on the performance in a multiple choice test conducted after the learning phase. Other studies also found effects on learning measures, arguing that the use of concept maps for navigation helps students to gain a better and also more durable knowledge (e.g. McDonald & Stevenson, 1999; Müller-Kalthoff & Möller, 2005). The effects of different navigation interfaces on search performance in a Web based learning environment were investigated by Carnot, Dunn, and Cañas (2001a, 2001b). Performance on finding relevant information for answering questions was better when using a concept map browser than when using alternative navigation interfaces (Web page, linked text).

Studies on the use of concept maps as navigation interfaces also analyzed navigation behavior. Generally, results show that navigation patterns are heavily influenced by the structure of the navigation tool (Boechler & Dawson, 2002). In the above mentioned study conducted by Hall et al. (2002) also differences in navigation behavior depending on the type of navigation interface (concept map vs. linear outline) were examined. They found that both interfaces constituted an anchor point in navigation. When considering the average time spent on the interface page, however, clear differences were identified. Those learners who had used the outline interface spent very little time on that page, whereas those with the concept map browser paused longer on the respective page. From this result, it appeared that learners in the concept map condition used the interface not only as an anchor but also as a study aid and a tool for integrating the to-be-learned information, whereas those using the outline interface simply went there for orientation.

Concept Mapping as a Means to Build Prerequisite Structures for Learning Objects and Competencies

Current developments in the context of e-learning are especially focusing on creating personalized and individualized learning experiences for each learner. Personalized learning aims at tailoring teaching to the needs, interests, and aptitudes of an individual. In personalizing learning it should be ensured that every learner achieves and reaches his/her highest standards possible. Usually, personalization proceeds by determining the current knowledge level of a learner, and probably also other individual characteristics or preferences (e.g. learning style), and utilizing this information for further teaching. For efficiently and adaptively uncovering the knowledge level of a learner, and for creating a personalized learning path, prerequisite structures on learning objects and assessment problems, or among competencies underlying them, are extremely useful. Prerequisite structures can, for example, build the basis for realizing adaptive navigation support, such as link annotation or link hiding, within an e-learning system (e.g. Brusilovsky, 1996, 2004). Knowledge Space Theory (Albert & Lukas, 1999; Doignon &

Falmagne, 1985, 1999; Falmagne, Koppen, Villano, Doignon, & Johannesen, 1990) provides a sound foundation for creating and realizing personalized learning experiences on the basis of prerequisite relationships (Albert, Hockemeyer, & Wesiak, 2002). In this context, concept maps can provide a useful instrument for determining prerequisite relationships and establishing prerequisite structures for learning objects, and competencies (Albert & Steiner 2005; Heller, Steiner, Hockemeyer, & Albert, 2006).

Let us take the knowledge domain of right triangles as an example to illustrate the application of concept maps in building a prerequisite structure among learning objects. Figure 5 depicts two typical learning objects (a) and (b), which actually are exercises. The knowledge necessary for solving these exercises can be identified as part of a concept map (see Figure 6). Exercise (a), for instance, requires applying the Pythagorean Theorem in order to calculate one side of a right triangle, given the other two sides. The relevant knowledge for this exercise therefore includes the concepts and relations in the concept map that describe the Pythagorean Theorem.

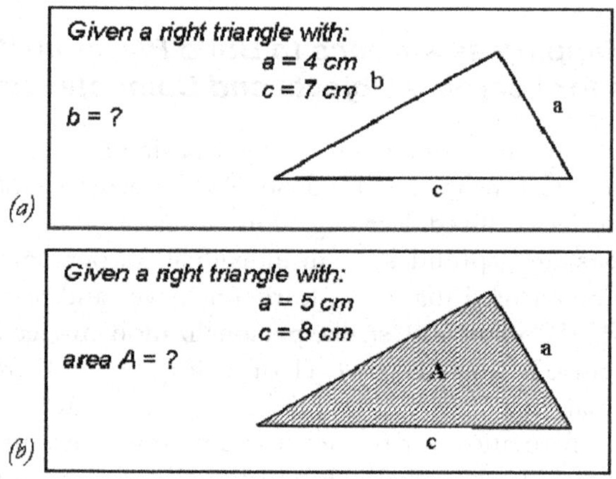

Figure 5: Example exercises (a) and (b) on the knowledge domain of right triangles (cp. Korossy, 1996, 1997)

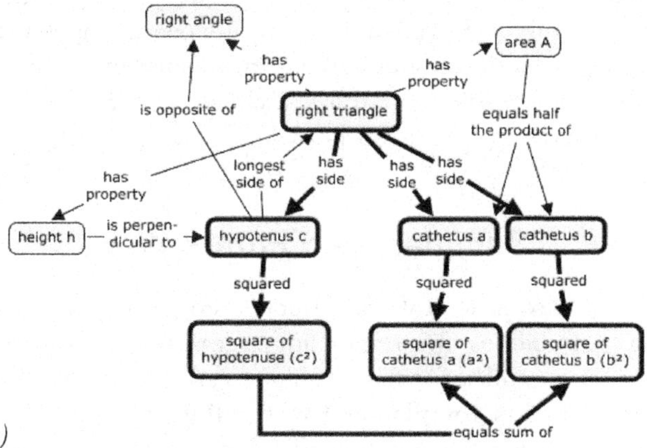

(a)

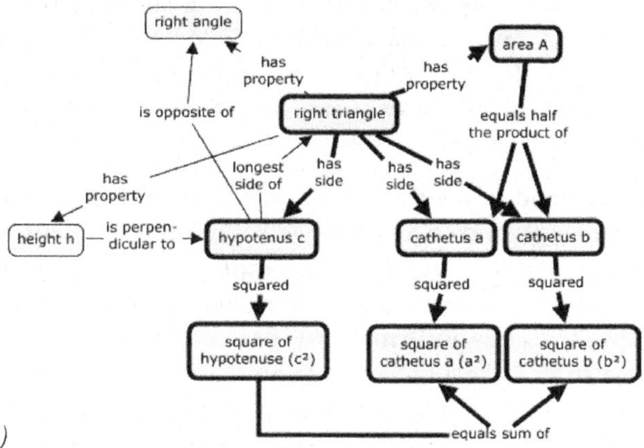

(b)

Figure 6: Concept maps with highlighted knowledge elements relevant for solving exerecises (a) and (b) (adapted from Steiner, 2004)

As can easily be derived from Figure 6, the representation of exercise (a) on the concept map constitutes a subset of that of exercise (b). Thus, it can be assumed that exercise (a) is a prerequisite to exercise (b). This information can be utilized for personalizing learning experience in e-learning. A learner having successfully mastered exercise (b) would be supposed to be also able to master exercise (a) and would therefore not be presented with it. On the other hand, it would not be useful to present exercise (b) to a learner who failed in mastering exercise (a).

In a similar manner as described above concept maps can be applied for deriving prerequisite relationships between competencies. In this case, competencies describing abilities and skills underlying observable behavior (e.g. being able to state the Pythagorean Theorem) are identified with the respective relevant knowledge elements of a concept map (Heller et al., 2006).

Summary and Future Work

Concept maps are powerful and effective cognitive tools that open up promising possibilities for organizing, representing, maintaining, and using knowledge and information. They can be beneficially applied in educational contexts for planning instruction, for presenting learning content, as a learning strategy for individual learners and groups, as well as for assessing knowledge and tracking changes in knowledge. In e-learning, concept mapping is able to provide support in course design and navigation. Concept maps can be utilized for deriving prerequisite structures among learning objects or competencies, which can then be implemented in e-learning systems for realizing an adaptive, personalized learning experience.

Current trends and developments in today's information and knowledge society in general, and in e-learning technology in particular, provide the potential of further and advanced applications of concept mapping. However, a few words of caution are in order, to. We have to be aware that the more or less informal way of 'traditional' concept mapping involves the risk of creating vague or ambiguous knowledge representations. Problems may result, for instance, from attempting to represent information elements that go beyond the expressive power of concept maps, as for example if-then rules. Reconciling a concept map as a formal knowledge representation therefore requires resolving this kind of ambiguities (Sowa, 2006).

Another field of current interest - which also provides the opportunity of applying concept maps - is that of so-called e-portfolios. An e-portfolio is a digitized collection of artifacts about an individual person (or, alternatively about a group, community etc.), providing evidence of the knowledge or competencies a person has acquired or developed over time (e.g. Lorenzo & Ittelson, 2005). They serve for documenting knowledge, skills, abilities, and learning, and for monitoring and evaluating performance and its development. In the context of career planning and job seeking, e-portfolios gain more and more importance. Commonly, e-portfolios are composed in natural language. An alternative possibility is their representation in form of concept maps. By documenting e-portfolios in concept maps that use a clearly defined vocabulary and a standardized grammar of representation, they could be made machine-readable. This would provide several advantages, such as making the e-portfolios of different people comparable. Furthermore, the description of an individual's knowledge or competencies contained in an e-portfolio could be compared with actual performance.

In sum, the present chapter was able to show that concept maps feature a broad field of applications in educational contexts in general, and in the scope of e-learning in particular. With today's wide availability of software tools for concept mapping this method is in a position to provide powerful support in planning, building and implementing e-learning. The use cases and benefits of concept mapping, however, are definitely not yet exhausted. Rather, current trends in information and learning technology are creating even further areas of application for concept mapping.

Acknowledgements

The work presented in this paper was supported by the European Community under the Information Society Technologies (IST) program of the 6th FP for RTD-project iClass (contract no. 507922). The authors are solely responsible for the content of this paper. It does not represent the opinion of the European Community, and the European Community is not responsible for any use that might be made of data appearing therein.

References

Abrams, R. (2000). *Meaningful learning: A collaborative literature review of concept mapping*. Retrieved June 2, 2005, from University of California Santa Cruz, Meaningful Learning Research Group Web site: http://www2.ucsc.edu/mlrg/clr-conceptmapping.html

Aebli, H. (1985). *Zwölf Grundformen des Lehrens [Twelve fundamental forms of teaching]*. Stuttgart: Klett-Cotta.

Afamasaga-Fuata'i, K. (2004). Concept maps & vee diagrams as tools for learning new mathematics topics. In A.J. Cañas, J.D. Novak, & F.M. González (Eds.), *Concept maps: Theory, methodology, technology.*

Proceedings of the First International Conference on Concept Mapping. Universidad Pública de Navarra, Pamplona. Retrieved February 15, 2006, from http://cmc.ihmc.us/papers/cmc2004-271.pdf

Aidman, E.V. & Egan, G. (1998). Academic assessment through computerized concept mapping: Validating a method of implicit map reconstruction. *International Journal of Instructional Media, 25*(3), 277-294.

Albert, D., Hockemeyer, C., & Wesiak, G. (2002). Current trends in eLearning based on knowledge space theory and cognitive psychology. *Psychologische Beiträge, 44*, 478-494.

Albert, D., & Lukas, J. (Eds.) (1999). *Knowledge spaces: Theories, empirical research, and applications.* Mahwah, NJ: Lawrence Erlbaum Associates.

Albert, D. & Steiner, C.M. (2005). Representing domain knowledge by concept caps: How to validate them? In T. Okamoto, D. Albert, T. Honda, & F.W. Hesse (Eds.), *The 2nd Joint Workshop of Cognition and Learning through Media-Communication for Advanced e-Learning* (pp. 169-174). Tokyo.

Allen, B.S., Hoffman, R.P., Kompella, J., & Sticht, T.G. (1993). Computer-based mapping for curriculum development. In M.R. Simonson (Ed.), *Proceedings of selected research and development presentations at the Annual Convention of the Association for Educational Communications and Technology.* New Orleans, LA: Research and Theory Division. (ERIC Document Reproduction Service No. ED 362 145)

Anderson, J.R. (1995). *Cognitive psychology and its implications* (2nd ed.). New York: W.H. Freeman and Company.

Anderson, J.R. & Bower, G.H. (1974). A propositional theory of recognition memory. *Memory & Cognition, 2*, 406-412.

Anderson-Inman, L. & Ditson, L. (1999). Computer-based concept mapping. A tool for negotiating meaning. *Learning & Leading with Technology, 26*(8), 6-13.

Anderson-Inman, L. & Zeitz, L. (1993). Computer-based concept mapping. Active studying for active learners. *The Computing Teacher, 21*(1), 6-11.

Ausubel, D.P. (1968). *Educational psychology: A cognitive view.* New York: Holt, Rinehart and Winston.

Bahr, G.S: & Dansereau, D.F. (2001). Bilingual knowledge maps (BiK-Maps) in second-language vocabulary learning. *The Journal of Experimental Education, 70*(1), 5-24.

Bahr, G.S. & Dansereau, D.F. (2004). Bilingual knowledge (BIK-) maps: Study strategy effects. In A.J. Cañas, J.D. Novak, & F.M. González (Eds.), *Concept maps: Theory, methodology, technology. Proceedings of the First International Conference on Concept Mapping. Universidad Pública de Navarra, Pamplona.* Retrieved February 15, 2006, from http://cmc.ihmc.us/papers/cmc2004-100.pdf

Bahr, G.S. & Dansereau, D.F. (2005). Bilingual knowledge maps (BiK Maps) as a presentation format: Delayed recall and training effects. *The Journal of Experimental Education, 73*(2), 101-118.

Baroody, A.J., & Bartels, B.H. (2000). Using concept maps to link mathematical ideas. *Mathematics Teaching in the Middle School, 5*, 604-609.

Bernd, H., Hippchen, T., Jüngst, K.-L. & Strittmatter, P. (2000). Durcharbeiten von Begriffsstrukturdarstellungen in unterrichtlichen und computergestützten Lernumgebungen [Working through concept structure representations in educational and computer based learning environments]. In H. Mandl & F. Fischer (Eds.), *Wissen sichtbar machen. Wissensmanagement mit Mapping-Techniken [Making knowledge visible. Knowledge management with mapping techniques]* (pp. 15-36). Göttingen: Hogrefe.

Boechler, P.M. & Dawson, M.R.W. (2002). Effects of navigation tool information on hypertext navigation behavior: A configural analysis of page-transition data. *Journal of Educational Multimedia and Hypermedia, 11*, 95-115.

Bolte, L.A. (1999). Enhancing and assessing preservice teachers' integration and expression of mathematical knowledge. *Journal of Mathematics Teacher Education, 2*, 167-185.

Brinkmann, A. (2003). Graphical knowledge display – mind mapping and concept mapping as efficient tools in mathematics education. *Mathematics Education Review, 16*, 25-48.

Brinkmann, A. (2005, October 25). Knowledge maps – tools for building structure in mathematics. *International Journal for Mathematics Teaching and Learning.* Retrieved March 15, 2006, from http://www.cimt.plymouth.ac.uk/journal/brinkmann.pdf

Bruhn, J., Fischer, F., Gräsel, C., & Mandl, H. (2000). Kooperatives Lernen mit Mapping-Techniken [Cooperative learning with mapping techniques]. In H. Mandl & F. Fischer (Eds.), *Wissen sichtbar machen. Wissensmanagement mit Mapping-Techniken [Making knowledge visible. Knowledge management with mapping techniques]* (pp. 119-133). Göttingen: Hogrefe.

Bruillard, E., & Baron, G.-L. (2000). Computer-based concept mapping: A review of a cognitive tool for students. In D. Benzie & D. Passey (Eds.), *Proceedings of the Conference on Educational Uses of Information and Communication Technologies* (pp. 331-338). Beijing: Publishing House of Electronics Industry.

Brusilovsky, P. (1996). Methods and techniques of adaptive hypermedia. *User Modeling and User Adapted Interaction, 6*, 87-129.

Brusilovsky, P. (2004). Adaptive navigation support: From adaptive hypermedia to the adaptive web and beyond. *PsychNology Journal, 2*, 7-23.

Cañas, A.J., Hill, G., Carff, R., Suri, J., Lott, J., Gómez, G., Eskridge, T.C., Arroyo, M., & Carvajal, R. (2004). CmapTools: A knowledge modeling and sharing environment. In A.J. Cañas, J.D. Novak, & F.M. González (Eds.), *Concept maps: Theory, methodology, technology. Proceedings of the First International Conference on Concept Mapping. Universidad Pública de Navarra, Pamplona.* Retrieved February 15, 2006 from http://cmc.ihmc.us/papers/cmc2004-283.pdf

Cardellini, L. (2004). Conceiving of concept maps to foster meaningful learning: An interview with Joseph. D. Novak. *Journal of Chemical Education, 81*, 1303 – 1308.

Carnot, M.J., Dunn, B., & Cañas, A. J. (2001a). Concept map-based vs. web page-based interfaces in search and browsing. *Proceedings of the 19th International Conference on Technology and Education, Tallahassee.* (ERIC Document Reproduction Service No. ED 462 979)

Carnot, M.J., Dunn, B., & Cañas, A. J. (2001b). *Concept maps vs. web pages for information searching and browsing.* Retrieved April 19, 2006, from http://www.ihmc.us/users/acanas/Publications/CMapsVSWebPagesExp1/CMapsVSWebPagesExp1.htm

Chang, K.E., Sung, Y.T., & Chen, S.F. (2001). Learning through computer-based concept mapping with scaffolding aid. *Journal of Computer Assisted Learning, 17*(1), 21-33.

Chang, K.-E., Sung, Y.-T., & Chen, I.-D. (2002). The effect of concept mapping to enhance text comprehension and summarization. *The Journal of Experimental Education, 71*(1), 5-23.

Chang, K.-E., Sung, Y.-T., Chang, R.-B., & Lin, S.-C. (2005). A new assessment for computer-based concept mapping. *Educational Technology & Society, 8*(3), 138-148.

Chiarello, C. (1994). Semantic memory. In V.S. Ramachandran (Ed.), *Encyclopedia of human behavior* (pp. 99-108). San Diego: Academic Press.

Chrobak, R. (2001). Metacognition and didactic tools in higher education. *Proceedings of the 2nd International Conference on Information Technology Based Higher Education and Training. Kumamoto, Japan.* Retrieved April 5, 2006, from http://www.eecs.kumamoto-u.ac.jp/ITHET01/proc/082.pdf

Clark, I.F. & James, P.R. (2004). Using concept maps to plan an introductory structural geology course. *Journal of Geoscience Education, 52,* 224-230.

Coffey, J.W. & Cañas, A.J. (2003). LEO: A learning environment organizer to support computer-mediated instruction. *Journal of Educational Technology Systems, 31*, 275-290.

Coffey, J.W., Carnot, M.J., Feltovich, P.J., Feltovich, J., Hoffman, R.R., Cañas, A.J., & Novak, J.D. (2003). A *summary of literature pertaining to the use of concept mapping techniques and technologies for education and performance support.* Retrieved June 6, 2005, from http://www.ihmc.us/users/acanas/Publications/ConceptMapLitReview/IHMC%20Literature%20Review%20on%20Concept%20Mapping.pdf

Collins, A., & Loftus, E. (1975). A spreading-activation theory of semantic processing. *Psychological Review, 82*, 407-428.

Collins, A., & Quillian, M. (1969). Retrieval time from semantic memory. *Journal of Verbal Learning and Verbal Behavior, 8*, 240-248.

Czerniak, C.M. & Haney, J.J. (1998). The effect of collaborative concept mapping on elementary preservice teachers' anxiety, efficacy, and achievement in physical science. *Journal of Science Teacher Education, 9*, 303-320.

Dabbagh, N. (2001). Concept mapping as a mindtool for critical thinking. *Journal of Computing in Teacher Education, 17*(2), 16-24.

Dansereau, D.F., Joe, G.W., Dees, S.M., & Simpson, D.D. (1996). Ethnicity and the effects of mapping-enhanced drug abuse counseling. *Active Behaviors, 21*, 363-376.

Denner, P.R. (1992). *Comparison of the effects of episodic mapping and traditional notetaking on the recall of historical text.* Paper presented at the Annual Meeting of the Northern Rocky Mountain Educational Research Association, Rapid City, October 7-10. (ERIC Document Reproduction Service No. ED 358 001)

De Simone, C., Schmid, R.F., & McEwen, L.A. (2001). Supporting the learning process with collaborative concept mapping using computer-based communication tools and processes. *Educational Research and Evaluation, 7*, 263-283.

Doignon, J.P. & Falmagne, J.C. (1985). Spaces for the assessment of knowledge. *International Journal of Man-Machine Studies, 23*, 175-196.

Doignon, J.P. & Falmagne, J.C. (1999). *Knowledge spaces.* Berlin: Springer.

Esiobu, G. & Soyibo, K. (1995). Effects of concept and vee mapping under three learning modes on students' cognitive achievement in ecology and genetics. *Journal of Research in Science Teaching, 32*, 971-995.

Falmagne, J.C., Koppen, M., Villano, M., Doignon, J.P., & Johannesen, L. (1990). Introduction to knowledge spaces: How to build, test, and search them. *Psychological Review, 97*, 201-224.

Greenwood, S., Phillimore, R., & Wang, Y. (2002). Supporting autonomous learners. *Teaching Forum (Oxford Brookes University internal publication), 50*, 21-26. Retrieved February 15, 2006, from http://www.brookes.ac.uk/virtual/NewTF/50/T50phillimore.pdf

Guastello, E.F., Beasley, T.M., & Sinatra, R.C. (2000). Content comprehension of low-achieving inner-city seventh graders. *Remedial and Special Education, 21*, 356-365.

Hall, R.H., Balestra, J., & Davis, M. (2000). *A navigational analysis of linear and non-linear hypermap interfaces.* Paper presented at the Annual Meeting of the American Educational Research Association, New Orleans, LA, April 2000. Retrieved May 28, 2005, from http://campus.umr.edu/lite/tech_reports/LITE-2000-02.pdf

Hall, R.H. & O'Donnell, A. (1996). Cognitive and affective outcomes of learning from knowledge maps. *Contemporary Educational Psychology, 21*, 94-101.

Hall, T., & Strangman, N. (2002). *Graphic organizers.* Retrieved February 15, 2006, from http://www.cast.org/publications/ncac/ncac_go.html

Heinze-Fry, J. (2004). Applications of concept mapping to undergraduate general education science courses. In A.J. Cañas, J.D. Novak, & F.M. González (Eds.), *Concept maps: Theory, methodology, technology. Proceedings of the First International Conference on Concept Mapping. Universidad Pública de Navarra, Pamplona.* Retrieved February 15, 2006, from http://cmc.ihmc.us/papers/cmc2004-033.pdf

Heller, J., Steiner, C., Hockemeyer, C., & Albert, D. (2006). Competence-based knowledge structures for personalised learning. *International Journal on E-Learning*, 5, 75-88.

Hibberd, R., Jones, A., & Morris, E. (2002). *The use of concept mapping as a means to promote and assess knowledge acquisition. Computers and Learning*

Research Group (CALRG) Report No. 202. Milton Keynes: The Open University, Institute of Educational Technology.

Jacobs-Lawson, J.M. & Hershey, D.A. (2002). Concept maps as an assessment tool in psychology courses. *Teaching of Psychology, 29*(1), 25-29.

Jegede, O.J., Alaiyemola, F., & Okebukola, P.A. (1990). The effect of concept mapping on students' anxiety and achievement in biology. *Journal of Research in Science Teaching, 27*, 951-960.

Jüngst, K.L. (1995). Studien zur didaktischen Nutzung von Concept Maps [Studies on the didactical use of concept maps]. *Unterrichtswissenschaft, 25,* 229-250.

Karasavvidis I. (2004). Concept mapping: Appropriation and resistance. In P.A.M. Kommers (Ed.), *Cognitive support for learning. Imaging the unknown* (pp. 241-251). Amsterdam: IOS Press.

Korossy, K. (1996). Kompetenz und Performanz beim Lösen von Geometrie–Aufgaben [Competence and performance on solving geometry problems]. *Zeitschrift für Experimentelle Psychologie*, 43, 279–318.

Korossy, K. (1997). Extending the theory of knowledge spaces: a competence-performance approach. *Zeitschrift für Psychologie, 205*, 53-82.

Lanzing, J. (2004). Everything you always wanted to know about … concept mapping. In P.A.M. Kommers (Ed.), *Cognitive support for learning. Imaging the unknown* (pp. 47-72). Amsterdam: IOS Press.

Lehman, J., Carter, C., & Kahle, J. (1985). Concept mapping, vee mapping and achievement: results of a field study with black high school students. *Journal of Research in Science Teaching, 22*, 663-673.

Lin, S., Strickland, J., Ray, B., & Denner, P. (2004). Computer-based concept mapping as a prewriting strategy for middle school students. *Meridian, 7.* Retrieved March 13, 2006, from http://www.ncsu.edu/meridian/sum2004/cbconceptmapping/

Lindsay, P.H. & Norman, D.A. (1977). *Human information processing. An introduction to psychology (2nd ed.).* Orlando: Adademic Press.

Liu, X. (2004). Using concept mapping for assessing and promoting relational conceptual change in science. *Science Education, 88*, 373-396.

Lorenzo, G. & Ittelson, J. (2005). *An overview of e-portfolios*. Retrieved April 24, 2006 from http://www.educause.edu/ir/library/pdf/ELI3001.pdf

Lumer, J. & Hesse, M. (2004). Concept mapping in the teaching of biology. In P.A.M. Kommers (Ed.), *Cognitive support for learning. Imaging the unknown* (pp. 81-91). Amsterdam: IOS Press.

Mandl, H. & Fischer, F. (2000). Mapping-Techniken und Begriffsnetze in Lern- und Kooperationsprozessen [Mapping techniques and concept maps in learning and cooperating processes]. In H. Mandl & F. Fischer (Eds.), *Wissen sichtbar machen. Wissensmanagement mit Mapping-Techniken [Making knowledge visible. Knowledge management with mapping techniques]* (pp. 3-12). Göttingen: Hogrefe.

Marchand, C., D'Ivernois, J.F., Assal, J.P., Slama, G., & Hivon, R. (2002). An analysis, using concept mapping, of diabetic patients' knowledge, before and after patient education. *Medical Teacher, 24*, 90-99.

Marshall, B., Chen, H., & Madhusudan, T. (2006). Matching knowledge elements in concept maps using a similarity flooding algorithm. *Decision Support Systems, 42*, 1290-1306.

Martin, D. J. (1994). Concept mapping as an aid to lesson planning: A longitudinal study. *Journal of Elementary Science Education, 6*(2), 11-30.

Martin, B.L., Mintzes, J.J., & Clavijo, I.E. (2000). Restructuring knowledge in biology: Cognitive processes and metacognitive reflections. *International Journal of Science Education, 22*, 303-323.

McClure, J.R., Sonak, B., & Suen, H.K. (1999). Concept map assessment of classroom learning: reliability, validity, and logistical practicality. *Journal of Research in Science Teaching, 36*, 475-492.

McDaniel, E., Roth, B., & Miller, M. (2005). Concept mapping as a tool for curriculum design. *Issues in Informing Science & Information Technology, 2*, 505-514.

McDonald, S., & Stevenson, R. (1999). Spatial versus conceptual maps as learning tools in hypertext. *Journal of Educational Multimedia and Hypermedia, 8*, 43–64.

McGowen, M. & Tall, D. (1999). Concept maps & schematic diagrams as devices for documenting the growth of mathematical knowledge. In O. Zaslavsky (Ed.), *Proceedings of PME23, Haifa, Israel, 3*, 281-288.

Mintzes, J.J., Wandersee, J.H., & Novak, J.D. (2001). Assessing understanding in biology. *Journal of Biological Education, 35*(3), 118-124.

Müller-Kalthoff, T. & Möller, J. (2005). Zum Effekt unterschiedlicher Navigationshilfen beim Lernen mit Hypertexten [On the effect of different navigation aids on learning from hypertexts]. *German Journal of Educational Psychology, 19*, 49-60.

Nicoll, G., Francisco, J., & Nakhleh, M.B. (2001). An investigation of the value of using concept maps in general chemistry. *Journal of Chemical Education, 78*, 1111-1117.

Novak, J.D. (1990). Concept maps and vee diagrams: two metacognitive tools to facilitate meaningful learning. *Instructional Science, 19*, 39-52.

Novak, J.D. (1998). *Learning, creating and using knowledge: Concept maps as facilitative tools in schools and corporations.* Mahwah: Lawrence Erlbaum.

Novak, J.D. (2001). *The theory underlying concept maps and how to construct them.* Retrieved May 10, 2005, from http://cmap.coginst.uwf.edu/info/printer.html

Novak, J.D. (2002). Meaningful learning: The essential factor for conceptual change in limited or inappropriate propositional hierarchies leading to empowerment of learners. *Science Education, 86*, 548-571.

Novak, J.D. (2003). The promise of new ideas and new technology for improving teaching and learning. *Cell Biology Education, 2*, 122-132.

Novak, J.D. (2004). Reflections on a half-century of thinking in science education and research: Implications from a twelve-year longitudinal study of children's learning. *Canadian Journal of Science, Mathematics, and Technology Education, 4*, 23-41.

Novak, J.D. & Gowin, D.B. (1984). *Learning how to learn.* New York: Cambridge University Press.

Novak, J.D. & Musonda, D. (1991). A twelve-year longitudinal study of science concept learning. *American Educational Research Journal, 28*(1), 117-153.

O'Donnell, A.M., Dansereau, D.F., & Hall, R.H. (2002). Knowledge maps as scaffolds for cognitive processing. *Educational Psychology Review, 14*, 71-86.

Patterson M, Dansereau D, & Wiegmann D. (1993) Receiving information during a cooperative episode: effects of communication aids and verbal ability. *Learning and Individual Differences, 5,* 1-11.

Pearsall, N.R., Skipper, J.E.J., & Mintzes, J.J. (1997). Knowledge restructuring in the life sciences: A longitudinal study of conceptual change in biology. *Science Education, 81*, 193-215.

Plotnick, E. (2001). Concept mapping: A graphical system for understanding the relationship between concepts. *Teacher Librarian, 28*(4), 42-44.

Quillian, M.R. (1966). *Semantic memory*. Cambridge: Bolt, Beranak, and Newman.

Reader, W., & Hammond, N. (1994). Computer-based tools to support learning from hypertext: Concept mapping tools and beyond. *Computer Education, 22,* 99-106.

Rebich, S. & Gautier, C. (2005). Concept mapping to reveal prior knowledge and conceptual change in a mock summit course on global climate change. *Journal of Geoscience Education, 53*, 355-365.

Rice, D.C., Ryan, J.M, & Samson, S.M. (1998). Using concept maps to assess student learning in the science classroom: Must different methods compete? *Journal of Research in Science Teaching, 35*, 1103-1127.

Roberts, L. (1999). Using concept maps to measure statistical understanding. *International Journal of Mathematical Education in Science and Technology, 30*, 707-717.

Ruiz-Primo, M.A. (2000). On the use of concept maps as an assessment tool in science: What we have learned so far. *Revista Electrónica de Investigación Educativa, 2.* Retrieved May 30, 2005, from http://redie.uabc.mx/vol2no1/contents-ruizpri.html

Ruiz-Primo, M.A. (2004). Examining concept maps as an assessment tool. In A.J. Cañas, J.D. Novak, & F.M. González (Eds.), *Concept maps: Theory, methodology, technology. Proceedings of the First International Conference on Concept Mapping. Universidad Pública de Navarra, Pamplona.* Retrieved April 3, 2006, from http://cmc.ihmc.us/papers/cmc2004-036.pdf

Ruiz-Primo, M.A., Schultz, S.E., Li, M., & Shavelson, R.J. (2001). Comparison of the reliability and validity of scores from two concept-mapping techniques. *Journal of Research in Science Teaching, 38*, 260-278.

Ruiz-Primo, M. A., & Shavelson, R. J. (1996). Problems and issues in the use of concept maps in science assessment. *Journal of Research in Science Teaching, 33*, 569-600.

Ruiz-Primo, M.A., Shavelson, R.J., Schultz, S. (1997). *On the validity of concept map-based assessment interpretations: An experiment testing the assumption of hierarchical concepts maps in science. CSE Technical Report 455.* Retrieved February 6, 2006 from http://www.cresst.org/Reports/ TECH455.pdf

Ruiz-Primo, M.A., & Shavelson, R.J., Li, M., & Schultz, S.E. (2001). On the validity of cognitive interpretations of scores from alternative concept-mapping techniques. *Educational Assessment, 7*(2), 99-141.

Rumelhart, D.E., Lindsay, P.H. & Norman, D.A. (1972). A process-model for long-term-memory. In E. Tulving & W. Donaldson (Eds.). *Organization of memory.* New York: Academic Press.

Rye, J.A., & Rubba, P.A. (2002). Scoring concept maps: an expert map-based scheme weighted for relationships. *School Science and Mathematics, 102*, 33-44.

Schau, C. & Mattern, N. (1997). Assessing students' connected understanding of statistical relationships. In I. Gal & J.B. Garfield (Eds.), *The assessment challenge in statistics education* (pp. 91-104). Netherlands: IOS Press.

Schau, C., Mattern, N., Zeilik, M., Teague, K. W., & Weber, R. (2001). Select-and-fill-in concept map scores as a measure of students' connected understanding of science. *Educational and Psychological Measurement, 61*, 136-158.

Schvaneveldt, R.W. (1990). *Pathfinder associative networks: Studies in knowledge organization.* Norwood: Ablex.

Shapiro, A. & Niederhauser D. (2004). Learning from hypertext: Research issues and findings. In D.H. Jonassen (Ed.), *Handbook of research on educational communications and technology* (pp. 605-620). Mahwah: Lawrence Erlbaum Associates.

Shavelson, R.J. & Ruiz-Primo, M.A. (2000). On the psychometrics of assessing science understanding. In J. Mintzes, J. Wandersee & J. Novak (Eds.), *Assessing science understanding* (pp. 303-339). San Diego: Academic Press.

Shavelson, R. J., Ruiz-Primo, M. A., & Wiley, E. W. (2005). Windows into the mind. *Higher Education, 49*, 413-430.

Sowa, J.F. (2006, April). *Conceptual maps.* Paper presented at the AERA Annual Meeting, San Francisco, California. Retrieved April 19, 2006 from http://www.jfsowa.com/talks/cmapping.pdf

Spector, J.M. (2006, April). *Expertise and assessing performance on complex and dynamic domains.* Paper presented at the AERA Annual Meeting, San Francisco, California.

Stanislaw, H. & Todorov, N. (1999). Calculation of signal detection theory measures. *Behavior Research Methods, Instruments, & Computers, 31*(1), 137-149.

Steiner, C.M. (2004). *Beziehung zwischen Concept Maps und Kompetenzstrukturen [Interrelation between concept maps and competence structures].* Unpublished diploma thesis, University of Graz.

Sternberg, R.J. (1999). *Cognitive psychology (2nd ed.).* Fort Worth: Harcourt Brace College Publishers.

Stoddart, T., Abrams, R., Gasper, E., & Canaday, D. (2000). Concept maps as assessment in science inquiry learning – a report of methodology. *International Journal of Science Education, 22*, 1221-1246.

Stoyanov, S. (1997). Cognitive mapping as a learning method in hypermedia design. *Journal of Interactive Learning Research, 8*, 309-323.

Stoyanov, S. & Kirschner, P. (2004). Expert concept mapping method for defining the characteristics of adaptive e-learning: ALFANET project case. *Educational Technology Research and Development, 52*, 41-56.

Sturm, J.M. & Rankin-Erickson, J.L. (2002). Effects of hand-drawn and computer-generated concept mapping on the expository writing of middle school students with learning disabilities. *Learning Disabilities Research & Practice, 17*, 124-139.

Swets, J.A. (1996). *Signal detection theory and ROC analysis in psychology and diagnostics. Collected papers.* Mahwah: Lawrence Erlbaum Associates.

Takeya, M., Sasaki, H., Nagaoka, K., & Yonezawa, N. (2004). A performance scoring method based on quantitative comparison of concept maps by a teacher and students. In A.J. Cañas, J.D. Novak, & F.M. González (Eds.), *Concept maps: Theory, methodology, technology. Proceedings of the First International Conference on Concept Mapping. Universidad Pública de Navarra, Pamplona.* Retrieved February 27, 2006, from http://cmc.ihmc.us/papers/cmc2004-260.pdf

Van Boxtel, C., Van der Linden, J., Roelofs, R., & Erkens, G. (2002). Collaborative concept mapping: Provoking and supporting meaningful discourse. *Theory Into Practice, 41*(1), 40-46.

Walker, J.M.T., & King, P.H. (2003). Concept mapping as a form of student assessment and instruction in the domain of bioengineering. *Journal of Engineering Education, 92*, 167-179.

Walker, J.M.T., King, P.H., & Cordray, D.S. (2003). The use of concept mapping as an alternative form of instruction and assessment in a capstone biomedical engineering design course. *Proceedings of the 2002 American Society for Engineering Education Annual Conference & Exposition.* Retrieved June 20, 2005 from http://www.vanth.org/mmedia/vanth0103/vanth0103cd/papers/WalkerConceptMap02. pdf

Wender, K.F. (1988). Semantische Netze als Bestandteil gedächtnispsychologischer Theorien [Semantic nets as elements of memory psychological theories]. In H. Mandl & H. Spada (Eds.), *Wissenspsychologie [Knowledge psychology]* (pp. 55-73). München: Psychologie Verlags Union.

West, D.C., Pomeroy, J.R., Park, J.K., Gerstenberger, E.A., & Sandoval, J. (2000). Critical thinking in graduate medical education. A role for concept mapping assessment? *Journal of the American Medical Association, 284*, 1105-1110.

Willerman, M. & Mac Harg, R.A. (1991). The concept map as an advance organizer. *Journal of Research in Science Teaching, 28*, 705-711.

Williams, C.G. (1998). Using concept maps to assess conceptual knowledge of function. *Journal for Research in Mathematics Education, 29*, 414-421.

Yin, Y., Vanides, J., Ruiz-Primo, M.A., Ayala, C.C., & Shavelson, R.J. (2005). Comparison of two concept-mapping techniques: Implications for scoring, interpretation, and use. *Journal of Research in Science Teaching, 42*, 166-184.

Zimmaro, D.M. & Cawley, J.M. (1998). *Concept map module.* Retrieved February 6, 2006, from Pennsylvania State University, Schreyer Institute for Innovation in Learning Web site: http://www.ttuhsc.edu/SOM/success/DHPS/Concept%20Map%20Module.htm

Authors

Christina M. Steiner completed her Diploma (MS) in psychology at the University of Graz (Austria). She is currently a researcher at the Cognitive Science Section (http://css.uni-graz.at) of University of Graz. Her work within several European R&D Projects on e-learning and Technology enhanced Learning focuses on the representation, modeling, and assessment of knowledge and competence. In particular, she is doing research on the use of concept maps as a means to build prerequisite structures among learning objects and competencies, and on the validation of concept maps. Further research interests include the evaluation of e-learning and the application of concept mapping as a learning and teaching strategy.

Dietrich Albert has a Diploma (MS) in Psychology from the University of Göttingen (Germany), and received his D.Sc. and his Habilitation in Psychology at the University of Marburg/Lahn (Germany) - before he became Professor of Experimental Psychology at the University of Heidelberg (Germany) for 16 years. Since 1993 he is Professor of Psychology at the University of Graz (Austria) and Head of the Cognitive Science Section (http://css.uni-graz.at) at the Department of Psychology working with psychologists, computer scientists and mathematicians. The academic year 2001/2002 he spent as a visiting professor at Hiroshima University, Japan. Currently he is the Chair of the 'Board of Trustees of the Center for Psychological Information and Documentation (ZPID)', Trier (Germany). His research topics cover several areas in cognitive psychology; his actual research focus is on knowledge and competency structures, especially competency-based theories of knowledge spaces, their applications, and empirical research. He and his team are currently involved in several European R&D Projects in e-learning and Technology enhanced Learning - especially in iClass - Intelligent Distributed Cognitive-based Open Learning System for Schools, ELeGI - European Learning Grid Infrastructure, ELEKTRA - Enhanced Learning Experience and Knowledge Transfer.

Jürgen Heller graduated from the University of Regensburg (Germany) in 1987 with a Diploma (MS) in Psychology. In 1991 he received his Ph.D. (Major: Psychology, Minor: Mathematics) and in 1998 his postdoctoral degree (Habilitation in Psychology), again from University of Regensburg. He was a visiting professor of psychology at the University of Erlangen-Nürnberg (Germany) during winter term 2000/2001, and at the University of Graz (Austria) during the academic year 2001/2002. From 2002 to 2004 he held an associate professorship at the University of Giessen (Germany). In 2004 he joined the Cognitive Science Section at the University of Graz (Austria) as a senior researcher and project coordinator of various European R&D Projects on e-learning and Technology enhanced Learning. Since 2006 he is holding a full professorship in methods of psychological research at the University of Tübingen (Germany). He is a member of the editorial board of the Journal of Mathematical Psychology. His research interests are mainly centered around psychological modeling and measurement in various areas of cognitive psychology and perception. Recent work addresses the representation and assessment of knowledge, with a particular focus on e-learning.

Buzzetto-More, N. A. (Ed.). (2007). *Advanced Principles of Effective e-Learning*
Santa Rosa, California: Informing Science Press. (pp. 113-131)

Chapter 4

Instructor and Student Roles in Two Opposing Pedagogies

Retta Sweat-Guy and Nicole Buzzetto-More

Introduction

Higher education is challenged to meet the needs of an increasingly diversifying student population. As a result, educators are looking at a variety of instructional strategies, many of which involve the use of technology and/or alternative delivery methods.

Advancements in technology have allowed institutions of higher learning to both enhance face-to-face instruction as well as deliver instruction beyond the traditional classroom (Rockwell, Schauer, Fritz, & Marx, 1999). At the fore is Web-based instruction, which has within the past decade had a significant and profound impact on the teaching and learning process.

Instructional strategies are methods or approaches used to achieve learning objectives. In recent years, there have been important new implications for how students learn best or how best to achieve learning objectives. It is important to empower students in the learning process in order to make them equally responsible for their own learning. This paradigm shifts the emphasis from content to a greater attention to process.

This chapter will focus specifically on instructor-centered and student-centered approaches to teaching and learning and how these strategies can be implemented in instruction as well as their relationship to learning theories. Additionally, a thorough literature review has been conducted to provide insight into the two approaches and a quantitative study has been presented which compares the two opposing pedagogies used in Web-based environments.

Literature Review

The literature is inundated with studies examining student achievement, student attitudes and perceptions, and student interaction in online learning environments; however, further research is needed to investigate how various instructional strategies impact student success in Web-based learning.

Huba and Freed (2000) cite the following as advantages to using a student-centered approach: (a) fosters collaboration; (b) involves students in the learning process; (c) integrates education and experience; (d) increase interaction between and among student and faculty; (e) provides prompt feedback; and (f) inclusive of diverse talents.

Berge (1999) recommends a student-centered approach for achieving higher order outcomes of instruction that are focused on analysis, synthesis, and evaluation. Land and Hannafin (1997) suggest that a student-centered approach provides "interactive, complementary activities that enable individuals to address unique learning interests and needs, study multiple levels of complexity, and deepen understanding" (p. 168). They conclude that the student-centered approach fosters a greater responsibility for learning and requires students to be more self-regulated in their approaches to learning. Additionally, they explain that the responsibility for learning is shifted to the student and that the instructor facilitates the learning by acting as a coach and resource guide. This results in the student becoming more proactive rather than reactive. Jonassen and Land (2000) explain that the student-centered approach lends itself to inquiry-oriented project-based learning activities, social discourse, and to alternative assessment regimes.

Northrup (2001) asserts that much of what exists online as web-based courses is instructor-centered citing that in much of the instruction, information is presented, followed by examples, simulated through practice, assessed through assignments and examinations, and in many cases, involves corrective feedback.

Phillips (2005) stresses the importance of feedback from educators as a crucial factor to student satisfaction in online learning experiences. He explains that online learning can lead to active learning that involves higher order thinking (e.g. analysis, synthesis, and evaluation) but explains that even in active learning the educator acts as a coach and facilitator that still must guide instruction, discussions, and activities. He refers to Gambesons' Seven Principles of Education for setting the foundation for active learning: 1) high expectations created by the instructor; 2) reciprocity and cooperation between learners; 3) active learning where the learning is involved in the learning process; 4) time on task; 5) feedback from educators that is timely, meaningful, and sustained; 6) student-educator interaction; and 7) respect for diversity.

In a study conducted using teacher education students, Eilon and Kliachko (2004) found that students encountered difficulty with student directed learning and that the instructor centered model led to greater student satisfaction. The teacher, they concluded, is crucial in providing guidance, mediation, management, technical and emotional support, and the understanding of course concepts.

Similarly, Stein, Wanstreet, Calvin, Overtoom, and Wheaton (2005) administered a survey and found that student satisfaction with online teaching environments correlated with the amount of interactions among students as well as between student and instructor. They concluded that increased interactions and instructor guidance led to greater course satisfaction and perceptions of knowledge gained.

In contrast, Schrum (2002), an advocate of the student-centered approach, found that flexibility in online courses allow students to be in control of their own learning which was found by Palloff and Pratt (1999) to make them more engaged.

Boyer (2004) whose study looked at adult learners explained that the student-centered approach allows learners the freedom to learn in a manner and at a pace that is aligned with their learning styles but also found in her study that instructor feedback and clearly defined expectations are necessary in order to provide clarification and ensure success in student-centered learning environments.

Whereas most studies that have looked at either the instructor's role in online education and/or the student verse instructor centered models, have focused on student satisfaction, this chapter will report the findings of a study that examined student learning outcomes in respect to the student verse instructor centered models.

Instructor-Centered

The instructor-centered approach to teaching and learning is historically the most frequently used instructional strategy (Tharp and Gallimore,

1988). The learning objective in an instructor-centered model is to transfer knowledge, information and/or skills from the instructor to the student. According to the theory underpinning this approach, instruction is developed and controlled by the instructor by providing declarative knowledge through lectures, textbook, and other resources (Schuh, 2004). In this environment, the focus is on content rather than student processing.

The control for learning is placed in the hands of the instructor; consequently, they decided what students will learn and how learning will take place. The instructor is considered to be the expert with respect to content knowledge; thus, using direct instruction to present the subject matter. The role of the instructor is to control the learning process by making decisions about what information to transmit and as well as the sequence for delivery. In short, the instructor-centered teaching focuses on the teacher as both authority and model.

With this approach, the role of the student is to acquire the information and demonstrate adequate knowledge acquisition. Therefore, activities based on the instructor-centered model would include presentations, lectures, and tests of knowledge.

Table 1 provides a summary of the instructor and student roles using the instructor-centered approach.

Table 1: Instructor and Student Roles in Instructor-Centered Approach

Instructor Role	Student Role
Knowledge is transmitted from the instructor to student	Students receive information
Instructor is expert on content knowledge	Emphasis is on right answers
Primary information giver	Only students are viewed as learners

In the instructor-centered approach, the instructor is the sole, infallible source of information, and the students memorize the information without questioning the correctness or origination of knowledge.

Student-Centered

The student-centered approach to teaching and learning is derived from Carl Rogers' Theory of Therapy, Personality and Interpersonal Relationships (Rogers, 1959). Rogers provides a basic hypothesis underlying person-centered teaching/learning as a constructive tendency characterized by realness, acceptance, and empathic understanding.

Motschnig-Pitrik and Mallich (2004) adopted Rogers' hypothesis to formulate the following statement:

> The better facilitators communicate and learners perceive the attitudes of realness, acceptance, and understanding and the more transparent the whole setting is, the deeper will be the learning processes at all three levels.

A student-centered approach to teaching and learning addresses the learner's intellect, social skills, personal experiences, and personality. Moreover, the student controls the learning process by determining what and how they will learn and by setting the pace of their own learning.

The student-centered approach focuses on the student by determining the student's needs and perceptions and by providing them with opportunities and possibilities to make choices about what they learn. Hence, decision-making, organization, and content are also determined by student needs where the prevailing activities include self-initiated projects, self-assessments, and performance or portfolio assessments.

The underlying philosophy of the student-centered approach to teaching and learning is that students learn best by receiving, interpreting, and discovering knowledge. In this environment, the role of the instructor is to provide learning opportunities and present possibilities for study, encouraging student exploration and discovery. The instructor serves as a coach and facilitator and is responsible for creating opportunities in which students:

- Set their own goals,
- Assume responsibility for their performance,
- Define strategies for their learning, and
- Identify indicators of success.

Additionally, the instructor coaches and mentors students in order to facilitate their learning and designs experiences that allow the student to acquire new knowledge and develop new skills.

The student's role is proactive. In a student-centered environment, students become reflective, creative, and practical thinkers, creating new meanings and relating such to past experiences. Students become self-guided and assume more responsibility for their learning.

Table 2 summaries the instructor and student roles in the student-centered approach.

Table 2: Instructor and Student Roles in Student-Centered Approach

Instructor Role	Student Role
Instructors serve as a Coach and facilitator	Construct knowledge; students are actively involved
Instructors and students learn together	Culture is cooperative, collaborative, and supportive
Provides students with opportunities and possibilities to make choices	Students set their own goals
Encourage student exploration and discovery	Discovers new knowledge

In the student-centered approach, "learners are treated as co-creators in the learning process, as individuals with ideas and issues that deserve attention and consideration" (McCombs and Whistler, 1997).

Learning Theories

When examining learning as a process, it is necessary to determine how learning takes place and which learning theories are present. Therefore, it is important to identify those learning theories that incorporate instructor-centered and/or student-centered pedagogy. Four theories incorporate at least one of the two opposing pedagogies: Bruner's Three-Form Theory, Gagne's Conditions of Learning, Merrill's Instructional Transaction Theory and Spiro's Cognitive Flexibility Theory.

Bruner Three-Form Theory

From Bruner's (1996) viewpoint, there are three ways in which individuals represent the real world. He ascertains that through action, icons and symbols, learners can convert reality into their own unique portrayal of reality. Bruner's theory illustrates action as enactment and demonstration, icons as images and pictures, and symbols as words and number. He defines the active form, which is based on stimulus-response theory, as the state of doing; the iconic form as a summary image or a mental picture of a path or pattern; and the symbolic form, which consists of language, as an abstract form since the word is a disconnection from the reality it represents. Bruner believes that teaching by telling is out of context of action and that learning by doing is a powerful way of learning.

Gagne Conditions of Learning

Gagne's (1985) theory is a systematic instruction and learning process in which he proposed a descriptive theory of knowledge consisting of five outcome categories: intellectual skills, cognitive strategies, verbal information, motor skills, and attitudes. Additionally, Gagne proposed a descriptive theory of strategy that consists of nine events of instruction: gaining attention, informing the learner of objectives and activating motivation, stimulating recall of prior knowledge, presenting the stimulus material, providing learning guidance, eliciting performance, providing feedback, assessing performance, and enhancing retention and transfer. From each outcome and event category, he then identifies the conditions necessary for learning to be efficient and effective, emphasizing that learning activities are the two-way transaction to stimulate learning capacity. Individual new information and knowledge creation is based on personal particular backgrounds and the results of two-way interaction between learners and instructor. These conditions of learning comprise Gagne's prescriptive theory of instruction.

Merrill Instructional Transaction Theory

Merrill, et al. (1996) proposed a theory to describe knowledge in terms of three types of knowledge objects: entities, activities, and processes. Also identified were six types of interrelationships among knowledge objects: components, properties, abstractions, associations between entitles activities, and processes. The instructional transaction theory also identifies a set of instructional algorithms that are called transaction shells, which consists of rules for selecting and sequencing knowledge objects. It also consists of sequence messages to knowledge objects which cause them to display a multimedia resource representing the knowledge object, display their name or description, change their location or property values, and consequently the multimedia resources associated with these change property values. The prescriptive theory consists of rules for selecting the transaction most effective for promoting learning of a given type of knowledge object or relationship between knowledge objects. Also the prescriptive theory indicates that interaction is a crucial factor that can assist learners to promote learning effectively with a given configuration of abilities and aptitudes.

Spiro Cognitive Flexibility Theory

Spiro's cognitive flexibility theory focuses on learning in the real world complexity and ill-structured knowledge domains (Spiro, Feltovich, Jacobson, and Coulson, 1991). From an ill-structured aspect of knowledge, advancing knowledge acquisition, such as attaining an understanding of important elements of conceptual complexity, the ability to use acquired concepts for reasoning and inference, and the ability to flexibly apply conceptual knowledge to novel situations, can be facilitated by the principles of this theory. This theory encourages a broader way of learning that emphasizes the importance of context in flexible learning environments. In support of this theory, Brown (1995) asserts that in order for a learner to fully comprehend the complexity and erratic variability of information, instruction must facilitate learning in complex, nonlinear domains. Based on the interactive view, Maddux, Johnson, and Willis (1997) argues that the non-linear nature can provide learners more opportunities to interact with instruction and with individual backgrounds, assisting learners to create their own knowledge.

Table 3 provides a summary of the learning theories presented and how it is represented in the learning process. As the table shows, these learning theories involve contrasting ideas as to the process of learning and the role that educators may take.

Table 3: Examining the Educator's Role in Learning Theories				
	Bruner	Gagne	Merrill	Spiro
Instructor-Centered			Arranges environment to elicit desired response	Structures content of learning activity
Student-centered	Works to establish communities of practice in which conversation and participation can occur	Facilitates development of the whole person		

Quantitative Study

This quantitative study contributes to the literature by examining the achievements of students using instructional strategies that have not yet been investigated in an online environment. Where most studies have focused on student satisfaction, this study examined student learning outcomes were in respect to the student verse instructor led models of instructions. Additionally, the study includes a sample population that is under represented in the literature relevant to distance learning experiences.

Participants

The participants in this study were undergraduate students within the Department of Business, Management and Accounting at the University of Maryland Eastern Shore. The University of Maryland Eastern Shore (UMES), a historically Black land-grant university, is a teaching, research and doctoral institution that represents a diverse student population that includes students from various social, economic and ethnic backgrounds.

All Business majors (Accounting, Business Administration and Business Education) are required to take the Business Communication course during their sophomore year. The study participants included those undergraduate students enrolled in the Business Communications course, BUED 333 sections 0101 and 0201 for the spring 2004 semester. They ranged in age from 18-25 and ethnicity background of participants were (90%) African American, (8%) Africans, (1%) Asian, and (1%) Caucasian. The number of participants in the instructor-centered and student-centered sections was 21 (11 male, 10 female) and 37 (19 male, 18 female), respectively.

Treatment

The Web-based course management tool, Web-CT, was used for course creation and delivery. The study examined two hybrid online sections of the Business Communications course offered during the spring, 2004. The instructional unit part 3, Correspondence, was administered across both treatment groups. Section 0101 of the Business Communications course was delivered using the instructor-centered instructional strategy while section 0201 was delivered using the student-centered instructional strategy.

In the instructor-centered section (0101), the delivery of unit content and activities were introduced and guided by the instructor. The instructor disseminated information, provided feedback, and corrective guidance. The e-mail tool was used to alert learners to the current state of activity and served to facilitate structure in the instructor-centered section. Students completed a continuous stream of activities that were time sensitive and which resulted in numerous opportunities for assessment and feedback.

The chat tool was used to deliver real-time lectures completely guided by the instructor. The discussion tool was used as a forum to hold threaded discussions on particular topics (1 per chapter). This was an instructor-moderated online discussion in which the students were given a topic to discuss with an established time period in which to respond. The content module included the following content-rich exercises and resources developed by South-Western Publishing: *E-mail it!* – consisted of online activities designed to teach students how to draft effective e-mail messages, negative messages, positive messages, and neutral messages. Students were able to submit drafted messages to the instructor for feedback and corrective guidance. *Glossary* – consisted of key terms with definitions arranged alphabetically for each chapter in the unit of study. *Self check* – included, for each chapter, a 10-question quiz to help students review material and a 5-question quiz using essay, short answer and other question types.

In the student-centered section (0201), learning was more self-directed where the learner was responsible for the pacing of learning activities. In this mode, the delivery of unit content and activities were inductive, experimental, and based on authentic and problem-based learning instructional concepts. The activities required students to analyze, synthesize, and evaluate. The methods, techniques, and communication media allowed students to work at their own pace and become self-directed and although due dates were provided; students were given long stretches for the completion of course activities. As a result, assessment and feedback was infrequent.

The Course Notes web page provided links to posted "lecture-type PowerPoint presentations" that were developed by South-Western Publishing. The chat tool was used as a platform to hold real-time concurrent dialog and conversation for analyzing case studies. Each case study (1 per chapter) offered in-depth examples of communication situations that related to chapter content, designed to challenge students to think critically. The students in this treatment group were also provided the self-check and glossary components as identified in the instructor-centered section. Additionally, the student-centered section was supported with expanded Internet activities made available by South-Western.

The instructional strategy provided the primary difference between the two treatment groups. The instructor-centered section was completed guided by the instructor for information dissemination, unit activities, and corrective feedback. The student-centered section was designed to increase the learners' responsibility for the outcomes of his or her own learning experience, allowing students the opportunity to learn the content at their own pace.

Instrumentation

The students were required to submit a Letter Portfolio. This assignment presented students the chance to practice writing several types of business correspondence discussed in the chapters designated for the study. The letter portfolio assignment included the following: (1) A Persuasive Memo, (2) A Letter that Asks a Question, (3) A Letter of

Appreciation, (4) A Letter of Sympathy, (5) A Letter Declining a Request for a Favor, (6) A Letter Rejecting a Job Applicant, and (7) A Persuasive Sales Letter from a Children's Toy Company.

Students were given an objective test to assess their understanding and knowledge of the learning objectives. The educational aims and objectives of the unit were to develop an understanding of how to use the message development framework to write both positive and neutral messages. The thirty test questions were selected from the test bank of the adopted textbook, which included multiple-choice questions on all learning objectives. An internal consistency reliability estimate (Cronbach's alpha) of .82 was obtained for the entire 30-item test. The test was administered during the face-to-face session immediately following the end of the unit. To prepare students for the objective test, an online self-check was available for each chapter, which provided immediate feedback.

Findings

The participants in this study were responsible for completing a unit of study using Web-CT. They were required to turn in a letter portfolio assignment and were administered an objective test at the end of the unit. The participants included 22 and 39 undergraduate students in the instructor-centered and student-centered sections, respectively. Before starting the web-based unit of study, all of the students received face-to-face training on how to use the various functions of Web-CT. As a follow-up to the face-to-face training, students in both groups were asked to complete the Online Technologies Self-Efficacy Scale (OTSES) to determine their level of confidence in using online technologies. They were then asked to determine if additional training would be necessary in completing the online unit of study. Those students requiring additional assistance were asked to schedule one-on-one training session(s) with the Web-CT coordinator.

The data collected for the study were analyzed using the Two-Sample T-Test to determine whether there were significant group differences between variables. For data analyses, the Statistical Package for Social sciences (SPSS) was used. For this study, the significance level was set at a=.05.

The first step in this analysis summarizes information about the variables (letter portfolio and the objective test) for this study. The averages, ranges and variances were calculated for each of the variables.

The total numbers of letter portfolios received from students in the instructor-centered and student-centered groups were 21 and 37 respectively. The mean score for the instructor-centered group was 82.57 with a standard deviation of 11.92. The scores for this group ranged from 56 to 100 with 75% of the participants receiving a score of 76 or higher. The mean score for the student-centered group was 83.37 with a standard deviation of 9.66. The scores for this group ranged from 69 to 98 with 75% of the participants receiving a score of 75 or higher. A low standard deviation for both groups suggests that the scores in each distribution are clustered around the mean. Outcome scores for the letter portfolio assignment were relatively high across both treatment groups (see Table 4).

Table 4: Means & Standard Deviations of Letter Portfolio Scores by Treatment

						Percentiles		
Treatment	*M*	*SD*	N	*Minimum*	*Maximum*	25	50	75
Instructor-centered	82.57	11.92	21	56.00	100.00	76	81	94
Student-centered	83.37	9.66	37	69.00	98.00	75	83	93

The descriptive statistics for the objective test scores by treatment are presented in Table 5. The total number of objective test scores for the instructor-centered and student-centered groups were 21 and 38, respectively. The mean and standard deviation for the instructor-centered group were (M= 81.80, SD = 15.22). The scores for this group ranged from 53 to 97 with 75% of the participants scoring 75 or higher. The mean and standard deviation for the student-centered group were (M = 75.26, SD = 17.00). The scores for this group ranged from 27 to 100 with 75% of the participants scoring 60 or higher. Similarly, as with the letter portfolio scores, the data suggest that the scores in each treatment group are clustered around the mean.

Table 5: Means & Standard Deviations of Objective Test Scores by Treatment						Percentiles		
Treatment	*M*	*SD*	N	*Minimum*	*Maximum*	25	50	75
Instructor-Centered	81.80	15.22	21	53.00	97.00	75	90	95
Student-centered	75.26	17.00	38	27.00	100.00	60	83	87

Descriptive statistics were calculated for all variables. Overall means and standard deviations for the letter portfolio and objective test are reported in Table 6.

Table 6: Descriptive Statistics fotr Letter Portfolio and Objective Test Scores						Percentiles		
Treatment	*M*	*SD*	N	*Minimum*	*Maximum*	25	50	75
Instructor-Centered	83.08	10.44	58	56	100	75	82	93
Student-centered	77.59	16.56	59	27	100	63	83	90

Data were generated from two primary sources, letter portfolio and objective test, to measure learner outcomes. This study investigates how learning strategies influence student performance in an online environment.

A two-sample t-test was performed to compare means for the distribution of continuous scaled variables (letter portfolio and objective test scores) between two independent treatment groups (instructor-centered and student-centered). The results of the analysis support the null hypothesis and claim no significant difference across treatment groups for the letter portfolio and objective test scores, $p>.05$ (see Table 7). It was concluded that students in both treatment groups achieved successful learning outcomes.

Table 7: Analysis of Mean Differences for Letter Portfolio and Objective Test Scores

		Levene's Test for Equality of Variances		t-test for Equality of Means				
		F	Sig	T	df	Sig (2-tailed)	Mean Diff	Std Error Diff
Objective Test	Equal variances assumed	.607	.439	1.468	57	.148	6.54	4.45
Letter Portfolio	Equal variances assumed	.415	.522	-.280	56	.780	-.80	2.87

This study contributes to the literature by examining the achievements of students using instructional strategies that have not yet been investigated in an online environment. Where most studies have focused on student satisfaction, this study examined student learning outcomes were in respect to the student verse instructor led models of instructions. Additionally, the study includes a sample population that is under represented in the literature relevant to distance learning experiences.

Descriptive statistics revealed that mean scores for the outcome measures (letter portfolio and objective test) were comparable across treatment groups. The primary data analysis revealed that the instructional strategy (instructor-centered vs. student-centered) had no effect on outcome as both groups were shown to perform the same with respect to the course project and objective test.

As higher education becomes increasingly more diversified, educators are increasingly challenged to make learning especially that which occurs online, a rewarding experience. It is paramount that researchers in the field strive to identify, all the variables relating to student, instructor, design, and cost that affect learning outcomes. This study acknowledged instructional strategy and the role of the instructor as a possible variable and examined the impact on student learning among minority learners.

Summary

The argument that 'one teaching and learning approach fits all', which is attributed to an instructor-centered approach, is under attack by educators and researchers alike. Diverse student needs, technological advances, and legislative mandates have prompted a paradigm shift in how instruction is presented.

A comparison of the two approaches, instructor- versus student-centered, reveals that the instructor demonstrates expertise in content knowledge. Similarly, instructors provide background data and content and pose questions that students can use to create meaning. Moreover, instructors acknowledge that students have different needs.

The way in which instruction is presented differentiates instructor-centered from the student-centered approach. In the instructor-centered approach, direct instruction is the predominant instructional method utilized. The instructor decides what is learned and how learning will take place. Students rarely construct their own meaning and assessment consists of objective tests.

In the student-centered approach, instructors encompass authority of the content knowledge but design flexibility for students to construct their own meaning. The focus is on how individual students learn. Students actively participate in the process by determining how and what they will learn.

Table 8 is a summary and comparison of instructor and student roles in the two opposing pedagogies. Both approaches have implications for instruction because they foresee different roles for learners and instructors and foresee different teaching materials.

Table 8: A Summary and Comparison of Instructor and Student Roles		
	Instructor Centered	Student Centered
Instructor's Role	Transmit information	Provide learning opportunities
Student's Role	Receive information and demonstrate competence	Consider learning opportunities and make final choices about what is o be learned
Activities/Assignments	Presentations, lectures, tests	Self-initiated projects

References

Berge, Z. L. (1999). Interaction in post-secondary Web-based learning. Education Technology, 39(1), 5-11.

Boyer, N. (2004). The learning contract process scaffolds for building social, self-directed learning. Quarterly Review of Distance Education. 4(4), 369-383

Eilon, B., & Kliachko, S. (2004). Perceptions of the teacher's roles by prospective elementary school science teachers in a Web-based biology course. Journal of Technology and Teacher Education. 12 (3), 339-360.

Huba, M. E., & Freed, J. E. (2000). Learner-centered assessment on college campuses: Shifting the focus from teaching to learning. Boston: Allyn & Bacon.

Jonassen, D. H., & Land, S. M. (2000). Theoretical foundations of learning environments. Mahway, NJ: Lawrence Erlbaum Associates.

Land, S. M., & Hannafin, M. J. (1997). Patterns of understanding with open-ended learning environments: A qualitative study. Education Technology Research and Development. Educational Technology Research and Development, 33(1), 47-73.

McCombs, B. L. (1997). Self-assessment and reflection tools for promoting teacher changes toward learner-centered practices. National Association of Secondary School Principals Bulletin 81, 1-14.

Moschnig-Pitrik, R. & Mallich, K. (2004). Effects of Person-Centered Attitudes on Professional and Social Competence in a Blended

Learning Paradigm. Educational Technology & Society, 7(4), 176-192.

Northrup, P. (2001). A Framework for Designing Interactivity into Web-Based Instruction. Educational Technology, 41(2), 31-39.

Palloff, R., & Pratt, K. (1999). Building learning communities in cyberspace: Effective strategies for the online classroom. San Fransisco: Jossey_Bass

Phillips, J. (2005). Strategies for active learning in online continuing education. The Journal of Continuing Education in Nursing, Mar/Apr 2005; 36, 2; 77

Rockwell, S. K., Schuaer, J., Fritz, S. M., & Marx, D. B. (1999). Incentives and obstacles influencing higher education faculty and administrators to teach. Online Journal of distance Learning Administration, 3(4).

Rogers, C. R. (1959). A Theory of Therapy, Personality, and Interpersonal Relationships as Developed in the Client-Centered Framework. In S. Koch (Ed.) Psychology: A Study of Science (Vol. 3), New York: McGraw-Hill, 184-256.

Schuh, K. L. (2004). Learner-centered principles in teacher-centered practices? Teaching and Teacher Education 20 (8), 833-846.

Schrum, L. (2002). Dimensions and strategies for online success: Vioces from experienced educators. Journal of Asynchronous Network Learning, 6(1), 57-67.

Stein, D., Wanstreet, C., Calvin, J., Overtoom, C., and Wheaton, J. (2005). Bridging the transactional distance gap in online learning environments. The American Journal of Distance Education. 19 (2). 105.

Tharp, R. and Gallimore, R. (1988). Rousing minds to life: Teaching, learning, and schooling in social context. New York, NY: Cambridge University Press.

Authors

Dr. Retta Sweat-Guy has an Ed.D. in Curriculum and Instruction with emphasis in Instructional Systems Design from the University of Kentucky. She received her masters in Public Administration from Kentucky State University and a Bachelor of Arts degree in Business Education from the University of Kentucky. Currently, she is Assistant Professor in the Department of Marketing and Business Education at Fayetteville State University in Fayetteville, North Carolina. Online teaching and learning serves as the basis for her current research for which she has presented at conferences and referred publications. She has and continues to conduct workshops and training sessions on the use of web-based course management tools.

Dr. Nicole A. Buzzetto-More is the coordinator of Business Education at the University of Maryland Eastern Shore; the Assessment coordinator for the Department of Business, Management, and Accounting; as well as the Co-Founder of the UMES Office of Instructional Technology. She received doctorate and masters degrees in communication and instructional technology from Columbia University and also holds degrees from the College of New Rochelle and Marist College. As a recognized assessment and e-learning expert, she is a frequent presenter at conferences across the globe. She has authored numerous publications in referred journals and has been recognized with an award from the American Distance Education Consortium.

Buzzetto-More, N. A. (Ed.). (2007). *Advanced Principles of Effective e-Learning*
Santa Rosa, California: Informing Science Press. (pp. 133-159)

Chapter 5

Bringing Activity into E-Learning – the Development of Online Active Learning and Training Environments

Claus Pahl

Introduction

Interaction is central in learning processes (Moore, 1992; Ohl, 2001). E-learning systems can act as mediators in the interaction between the learner and her or his environment, i.e. content, peers, and instructors. Engaging the learner in a learning process through interaction is a central element of successful learning design (Sims, 1997). The current predominant focus on knowledge transfer in e-learning is partly a result of a lack of interactive multimedia technologies. With the recognition of skills training as being equally important to knowledge acquisition, more work has been done on activity-based learning and training supported by interactive multimedia technology (Trikic, 2001). With advances in multimedia and Web technologies, a shift from purely knowledge-based learning towards activity-based learning and training can be observed (Okamoto et al., 2001).

Interactive Web and multimedia technologies support skills-oriented training in learning technology systems in addition to knowledge-based learning. In particular, in e-learning environments, the learner-content interaction is often more central than the learner's interaction with instructors and peers (Ohl, 2001). The focus here is on the development of active learning objects based on interactive educational multimedia. Advanced uses of multimedia, in particular interactive educational multimedia, is a technical solution that can support activity-based e-learning and e-training (Elsom-Cook, 2001). Active learning is characterized by knowledge- or skills-level interaction of the learner with e-learning system and content, i.e. goes beyond the typical navigation and delivery management interaction. The input of the learner in interactions is meaningful in the context of the subject. The interactions between learners and instructors or collaborations between learners are neglected here.

Current approaches for knowledge transfer-oriented learning and their development do not often apply in this context. Support frameworks for multimedia development for e-learning environments exist (Heller et al., 2001). However, the focus of these frameworks is mainly on knowledge acquisition. Understanding the requirements of activity-based learning and training and techniques to develop interactive educational multimedia are therefore needed. Both developers (for instruction and software) and instructors need support for the development and deployment of this type of learning and the supporting technology.

This chapter presents an experience with the development of online active learning and training applications within a constructive framework. The aim is to guide the development of an activity-centered e-learning or e-training system for developers and instructors or the selection process of a suitable system needed by an instructor. The role of the instructor in the development process is explained. The focus is on interactive content development, based on a sound learning design. Those aspects that distinguish the development of a media-based learning object for active learning and training from classical knowledge-oriented learning and learning objects development are emphasized. The framework is illustrated using a third-level Web-based computing course in which various learning and training activities ranging from controlled animations to graphical modeling and text-based programming activities are integrated in a realistic setting. This system, a database learning environment called the Interactive Database Learning Environment IDLE, is a typical example of a wide range of online learning systems that support activities through animations, simulations, and other forms of processing.

Literature Review

The focus of this chapter is on activity-based learning and training. Active learning plays an important role in recent instructional design approaches. Learning is not an activity. It is a cognitive process that is rather a by-product of understanding (Mayes & Fowler, 1999). This cognitive dimension cannot be directly addressed using a software tool.

Essential elements of the learning experience, like direct interactions between learner and instructor or other learners, are missing. Activity theory can explain the role of the e-learning system in this context. Learning can be seen as a process in which learners actively construct knowledge and acquire skills. In the context of computer-supported interactive learning and training environments, the role of technology is that of a tool that mediates the interaction between learner and content. Activity theory is a conceptual framework that can describe the structure, development, and context of computer-supported learning activities (Nardi, 1997).

The emphasis on the interaction between learners and their environment explains the principle of tool mediation. Tools shape the way humans interact with reality. Educational tools reflect experiences learners and instructors have made in trying to solve particular problems. The learner should be engaged in solving meaningful problems in an activity-based, realistic setting.

Learning should be an active process in which interactivity is central (Northrup, 2001). Moore (1992) distinguishes three types of interactivity – learner-learner, learner-instructor, and learner-content interactions. It is often argued (Sims, 1997; Ohl, 2001) that content has a more central function in computer-based education than interaction with peers or instructors. The focus here is on learner-content interactions, where content is provided in the form of interactive multimedia. Learner-content interaction in computer-supported learning and training actually occurs as interaction with interactive multimedia features (Boyle, 1997). Educational multimedia systems are usually hypermedia systems providing structure through hierarchy and guidance for learning tasks through navigation topologies (Jonassen & Mandl, 1990). Crucial for educational multimedia are the multimedia interface and the interaction dialogues a multimedia system allows through channels (text, mouse, etc.) and languages (natural, formal, etc.).

The virtual apprenticeship model (Murray et al., 2003) is a pedagogical theory – based on the activity model – that defines an activity-based and skills-oriented learning and training framework for the IDLE system. An apprentice is a learner who is coached by a master to perform a specific task. In an e-learning and training environment, the master's role is often replaced by an intelligent software tool such as IDLE. The apprenticeship model determines a number of aspects including the activity purpose and the degree of involvement, interaction styles (e.g.

the organization of learning into sessions and cycles), and the interconnectedness of activities and features. The virtual apprenticeship model puts an emphasis on skills-oriented activities with a high degree of involvement of the learner.

Usability is central in user-oriented systems that support active learning. Technical and pedagogical usability can be distinguished as two aspects of usability (Melis et al., 2003). Technical usability aims at enabling trouble-free interaction of the user with the system. Pedagogical usability facilitates the learning process. In the context of e-learning systems, technical usability should enhance the pedagogical usability. Usability in both forms needs to be a central design and also evaluation focused on both developers and instructors.

The development of an e-learning system is a software development process that is heavily influenced by the specific context of the application domain, e-learning. Some process models to guide and structure the development process have already been proposed. Heller et al. (2001) have focused on the multimedia perspective, which is important for learning content development. Virvou and Tsiriga (2001) have used object-oriented software development, a very common software engineering methodology, to develop e-learning systems. Boyle (2003) follows a similar direction by combining object-oriented development methods with pedagogical principles for his development process model. None of the existing approaches, however, has addressed the specifics of active learning. In order to make a software process model accessible for domain experts, the process model also needs to be embedded into a learning-specific conceptual framework. The development of these systems requires the joint participation of software developers and instructors as domain expert in this process

Case Study

IDLE is the Interactive Database Learning Environment – a Web-based learning and training system for database modeling and programming. IDLE is based on the virtual apprenticeship model, which emphasizes skills-oriented learning activities. The system acts as a mediator between a learner and interactive content in a realistic training environment. This case study aims to illustrate the importance of a systematic development approach for e-learning systems development and deployment and focuses on aspects that are often overlooked.

Various forms of learning and training activities and how they can be supported by interactive multimedia are illustrated.

IDLE Overview

An interactive learning and training environment for SQL database programming is embedded into the online courseware system IDLE. Focusing on the SQL element illustrates the interactivity aspects. The SQL part forms a central part of this course as database programming is one of the core learning objectives of the course. Programming is a skill that needs to be acquired by the learner. Moreover, this course is also an introduction to database engineering. Therefore, understanding and mastering the overall development process of a database application is equally important. Database programming in SQL also requires conceptual understanding of the underlying data model with its structures, operations, and constraints.

Solutions to programming problems, which are presented as a guided tour through the material, can be submitted through a Web interface to a remote database server, which executes the input and replies with data from a database, or error messages. Scaffolding in form of feedback, self-assessment functionality, and links to background material is available. The tutorial prepares the learner for coursework, such as lab tests and projects, and final exams. The IDLE system aims at providing the learner with a realistic learning context by integrating features and problems into a learning environment that are similar to tools and tasks that would be faced by a database engineer in a real development scenario.

Classification of IDLE Features

Four different features for the different aspects of database programming are provided:

- Conceptual knowledge. Conceptual knowledge is presented in a virtual lecture system. The learners use an audio-visual presentation that presents the conceptual background. Recorded speech of the lecturer is synchronized with material in overhead form. Learners can control this lecture-style presentation through the usual interactive features of an audio player.

- Procedural knowledge. SQL is about the execution of instructions. Procedural knowledge is presented in an animated tutorial system. SQL is a language that implements different database operations. An animated tutorial using flash animations illustrates the execution of these operations is a step-by-step fashion using examples. Learners can execute these operations in small steps. Animated tutorials are particularly useful to illustrate operations that are sometimes conceptually difficult to understand.

- Programming skills. SQL programming is the core activity, supported by an interactive tutorial that guides the learner through exercises to be worked on within the system. SQL queries are often complex and difficult to formulate. Queries are supported through an interactive tutorial. The tutorial guides a learner through a sequence of exercises with increasing difficulty. Each unit addresses a particular aspect of SQL querying. The feature provides an input interface for each exercise where a learner can type in an SQL solution and submit this solution to a remote database server that executes the query and returns the result. Syntactic and semantic feedback is available. This feature provides links to the relevant background material (conceptual and procedural knowledge features).

- Development skills. SQL programming is part of the overall database application development process, which supported by an integrated lab environment with modeling, programming, and analysis features. The development of a database application is a multi-stage process including the stages modeling, programming, and analysis and optimization. The database course environment provides interactive, integrated lab features for all three activities. The learner is provided with a workspace in which s/he can create and store a data model, which is interconnected with the other features. An integrated, realistic lab environment that resembles features of database development environments is the central feature.

Conceptual understanding of principles and concepts is of course required before practical work can start. However, the aim of the tutorial system is to allow learners to go quickly into the practical features by supporting a learning-by-discovery style, allowing them to acquire skills, but also to construct and deepen their conceptual knowledge through activities in meaningful and realistic problems. Consequently, the practical features are well linked to the respective background.

An Interaction Space – Activity and Interaction in Learning and Training

This design section introduces an activity and interaction design framework – called an interaction space – focusing on learning activities and training. Mapping learning and activity types onto the most suitable media type is an essential step that will be illustrated by the IDLE implementation. Central is a classification of learner activities based on various aspects such as learning purpose (knowledge or skills) or degree of involvement. The classification is illustrated by different activity types supported within IDLE. This section focuses on early design stages, drawing attention to some aspects that are sometimes ignored, before classical learning design instruments would be deployed. This is a stage where in particular instructors are highly involved as domain experts.

Active Learning

In IDLE, database application development provides a meaningful problem that requires a learner to develop and deploy a database application with its structural and operational elements within the learning environment. The database courseware system creates a realistic setting by integrating tools into a learning and training environment that resemble tools of a real database development environment. Modeling, programming, and analysis features interacting with an enterprise-scale database server can provide a realistic setting. IDLE is a software tool that facilitates these activities in a guided learning process. Learners learn to solve problems in a dialogue with the system.

The following instructional guidelines, developed by domain experts, have served as requirements for the IDLE system development in order to enable active learning and training in a pedagogical framework defined by activity theory:

- The active participation of the learner is essential.
- Active construction of knowledge and skills results in an increased ownership of the learner in the learning process.
- Meaningful projects allow learners to acquire skills and experience in database programming and development.

- A realistic setting improves the learning experience and demonstrates the applicability of knowledge and skills.
- Guidance and feedback provide instructional support in the environment.

Activities such as programming are at the centre of the IDLE learning and training strategy. However, supporting the learner through scaffolding, e.g. guidance and feedback, is equally essential from the instructional perspective (McLoughlin et al., 2000). In addition to mediating between learner and database tools, the environment must fulfill functions of the instructor. The environment needs to replace central tasks of the instructor in form of a virtual master that guides a learner through exercises and that provides immediate feedback on activities. Again, instructors can provide crucial input here. Each learning and training activity needs to be complemented by links to the background (conceptual and procedural knowledge in form of virtual lectures and animated tutorials) relevant and problem-related for the activity in question.

Interaction Model

The learning and training activities facilitated by educational multimedia interactions between learner and content is captured in form of an interaction model for learning activities. The aim of this interaction model is to support the design of educationally sound interactive learning activities. A taxonomy of activity types is a central element of this interaction model that helps to categorize activities at an early stage. The categorization provides the developer and instructor with pointers to best practice, since each category is usually implemented in a specific way. Two aspects of activity types (purpose and involvement) that help to describe activities can be distinguished.

Three activity types can be defined based on the purpose of the learning process:

- Declarative knowledge acquisition activities: the aim is the acquisition of declarative knowledge in order to reason about it.
- Procedural knowledge acquisition activities: the aim is the acquisition of procedural knowledge in order to reason about it.

- Skills acquisition activities: the aim is the acquisition of procedural knowledge and experience in order to perform the instructions.

The second category is important in particular in the sciences and engineering domain where an understanding of the subject activities is required for a learner.

The style of the activity execution can be characterized based on the degree of involvement and influence of the learner on the environment. Types ranging from system-controlled to learner-controlled environments can be distinguished:

- Observation: a form of knowledge acquisition with no influence on the environment activities by a passive learner.
- Controlling: a form of knowledge acquisition mixed with knowledge production, based on observational elements, but allowing the learner to influence the environment activities to control their ordering.
- Creation: a form of activity where knowledge or skills are created by producing some form of artifact that can be processed by the learning environment.

The individual types for each of the categories are not meant to be exclusive. A more fine-granular classification can replace the above types if needed. Often the two aspects of activity types are related. Declarative knowledge is often acquired through observation, procedural knowledge for reasoning purposes through controlled animations, and skills through act creation and processing. The learning-by-doing idea is part of the IDLE active learning approach. It captures the interplay of knowledge acquisition and knowledge creation in an interactive process with the learning environment. This focus is widened in IDLE by considering knowledge acquisition on the one hand and skills and experience acquisition on the other hand as dual sides of learning and training.

In an interaction model, the activity model is the central element that defines content interactions, but which is complemented by two other aspects: goals, which are meta-level descriptions that define the learning objectives of an activity, and knowledge objects, which capture the knowledge underlying each activity in declarative and procedural form.

The ultimate objective at this development stage is the identification and the design of interactive learning objects that implement the activities in question and allow the learner to achieve the goals defined (Boyle, 2003). The interaction model – consisting of activities, their goals, and the underlying knowledge objects – helps the instructor and content developer to gather the required information.

Choosing the virtual apprenticeship approach as the underlying pedagogical framework is the first main IDLE design decision, which defines the context for the interaction model. Based on this decision, the characterization of individual activities, like the SQL programming described above, can start.

The main IDLE activity categories in terms of the interaction model are summarized in Table 1. Further categorization is necessary for a detailed design. For instance, the lab activity could be refined into specific activities such as graphical design, programming, or optimization. Learning object identification is the central objective at this stage. The aim is representing each activity in Table 1 by a composite learning object. This illustrates the need to involve instructors in this process. Within each of these composite learning objects, individual input will address specific topics. One of these specific activities is illustrated now, looking in particular at the interactions between learner and learning objects and at the composition (sequencing) of smaller activities.

Table 1. Some IDLE Activities and their Types based on Learning Purpose and Degree of Involvement

Activity	*Activity Type (Purpose)*	*Activity Type (Involvement)*
lecture participation	declarative knowledge acquisition	Observation
tutorial participation	procedural knowledge acquisition	Controlling
lab participation	skills acquisition	Creation

One of the skills acquisition activities is SQL database programming, which has the goal of equipping learners with the ability to formulate complex database queries. Integrated with a database system, the learner – a virtual apprentice – works through guided material covering a range of individual problems. Each problem is based on a submission- and execution-cycle with a high degree of involvement of the learner through knowledge creation. Each solution – content-specific knowledge that is created by the learner – is analyzed and, based on an individual activity history and integrated assessments, personalized feedback is given by a virtual master. At this level, the concern is the abstract classification of learner activities from the interaction model in order to design the learning object. For the database course IDLE the central design decision at this level is to focus on an integrated approach with a strong support of skills training activities.

An Interaction Infrastructure – Interactive Educational Multimedia

This implementation section discusses the technical aspects of interactive educational multimedia in the context of interaction for activity-based learning and training. Technical notions such as channels and interface languages shall be introduced as central aspects of educational multimedia that explain interactions between learners and content. Designing active learning and training in terms of abstract media concepts is central for the correct and successful implementation of the learning design from previous stages.

Interactive Multimedia

Interactive multimedia for activity-based learning and training can be distinguished into interaction with knowledge media and with activity media. Activity-based training focuses on skills-oriented activities, but needs to be integrated with knowledge learning aspects. Knowledge media focus on knowledge information that can be communicated. Activity media focus on knowledge-based acts that are produced and processed in activities. The purpose of interactive educational multimedia is twofold:

- In addition to knowledge-level interaction, domain-specific activities need to be facilitated; activity-level interaction with educational multimedia feature through acts and instructions has to be enabled.
- The instructor can be replaced by a virtual form of an intelligent educational multimedia feature that provides advice and feedback, adding more meaning to the interaction.

Educational multimedia can be classified through different metadata facets:

- Channels are abstractions of a communication device, characterized by modality.
- Languages enable the encoding of information in a common notation for the communication over a channel.
- The activity purpose distinguishes whether declarative knowledge reasoning, procedural knowledge reasoning, or skills acquisition is the aim.
- The activity style allows the classification of activities into observation, controlling, and creation, which describes the degree of influence of a learner on the environment.
- The content topic is the topic or domain within which activities or knowledge-level access is provided.

Learning objects have been used as a design notion in the previous section. These learning object designs and their underlying activities need to be implemented, which is mainly the work of a software or content developer. The focus here is on learning objects that realize learning activities as learner-content interactions. In particular, the media to implement activities are of central importance. The term interactive educational multimedia (IEMM) shall denote media types especially suited to implement active learning objects. A number of standard mappings between activities and media types have emerged:

- Lecture participation (aiming at declarative knowledge acquisition) is often suitably implemented by audio-visual presentations.
- Tutorial participation (aiming at procedural knowledge acquisition) is often suitably implemented by animations.
- Lab participation (aiming at skills acquisition) is often suitably implemented by simulations.

Specific media allow for the appropriate level and degree of interaction. Determining the most suitable media type and the interaction with the medium is the central activity at this stage. The following classification scheme and the investigation into interaction channels in the remainder of this section support this process to see a learning object as an interactive multimedia object.

IDLE supports three classical forms of third-level teaching, lectures, tutorials, and labs, in a virtual format. These three forms can be described using the facets of the educational multimedia classification scheme – see Table 2, which shows how some selected learning activity styles for particular topics are mapped onto multimedia features. For example, a simulation can be a subcategory of a moving pictures/images language. However, the elements of simulations can be identified and have meaning in the context of content (e.g. tables or records in the database context). Equally, operations (simulation activities) are represented as procedural knowledge.

Table 2. Sample IDLE Media Classification

Facet Activity	*Channel*	*Language*	*Purpose*	*Type*	*Knowledge Object*
lecture	text and audio	natural language	declarative knowledge	observation	introduction to databases
tutorial	dynamic animation	simulation	procedural knowledge	controlling	relational algebra
Lab	text	formal language	skills training	Creation	SQL

Interaction Channels

Multimedia is about channels and meaningful communication along these channels. Often, a natural language such as English is used over a text channel (written English) or over an audio channel (spoken English). For this context, a number of education-specific channels can be identified – supporting partly more formal languages, partly languages specific to the subject or instruction context. Two types of channels can be distinguished – those that support core content-oriented learning activities and those that are part of the meta-context of instruction; the latter including instruction-related learner actions and coaching actions by a master or instructor.

- Declarative knowledge (supporting core activity): declarative knowledge usually communicated in a domain-specific natural or formal language.
- Procedural knowledge (supporting core activity): procedural knowledge usually communicated in a domain-specific natural or formal language.
- Skills (supporting core activity): acts to be processed in form of activities are communicated with corresponding execution instructions.
- Actions (supporting meta activity): instruction-related actions executed by the learner, typically navigation or location of learning objects.
- Feedback (supporting meta activity): response of the system for each core activity.
- Coaching (supporting meta activity): meta-level information capturing an instructor's advice and guidance.

Multimedia interface languages capture and constrain channel communications. A language defines interaction dialogues; it describes the legal actions, how a learner can engage in an activity or how a learner can perform a task towards a learning goal.

The IDLE channel and language characterization in Table 2 is high-level. These two aspects can be described in more detail. Table 3 provides a channel-oriented view on IDLE, listing the educational channel types and some sample features that are based on these channels.

Table 3. Sample IDLE Media Channels

Channel	*Feature*	*Activity*	*Language*
declarative knowledge	database introduction lecture	HMTL and audio-based synchronized virtual lecture	natural language (written and spoken)
procedural knowledge	relational algebra animation	interactive simulation of algebra operator execution	formal language (interaction – animation control)

Channel	*Feature*	*Activity*	*Language*
skills	SQL programming lab	submission of query solutions and dynamic page update by system	formal language – SQL (solution and result)
action	SQL tutorial navigation	guided tour through a series of connected exercises	formal language (interaction – navigation)
feedback	SQL programming lab	correction and provision of partial solutions for exercises	semi-formal language (error classification)
coaching	self-assessment	multiple choice questions and virtual master's feedback	natural language (written)

Interaction Specification

Languages for the educational context can be classified based on content-related aspects:

- Natural languages – in text or audio form – are often the basis of content.
- Formal languages – in text form – are often involved if some form of mechanical computerized processing is part of the subject domain.
- Simulations – automated execution of some real-world activities – are based on objects and procedures from the subject domain.

On the most basic level the learner interacts with multimedia usually through keyboard and mouse; output can be static visual (text, graphics), dynamic visual (animations, video), or involving other modalities such as audio. The basic inputs are part of low-level activities such as navigation (knowledge acquisition request) or text input/submission (knowledge generation). A learning activity can be composed of more basic activities. The dialogue part of an interaction language consists of:

- basic activities like select (knowledge acquisition by learner), submit (knowledge generation by learner), reply (response to knowledge acquisition/generation);

- activity combinators like sequence, iteration, choice;
- system components such as learner and multimedia learning objects in an e-learning architecture.

An example shall illustrate how this language is used to express abstract interaction behavior between learners and learning objects. The expression

iterate
sequence of
LR.*select(exercise)*
LR.*submit(solution)*
LO.*reply(result)*

is the interaction specification of an exercise activity scenario. A language needs to facilitate declarative and procedural knowledge communication, skills-oriented activity execution, learner actions, and meta-level pedagogical interactions (coaching). The select activity denotes a learner action; submit and reply support skills-oriented activities; reply could, in addition to results for e.g. SQL submissions, also convey meta-level feedback and coaching. This specification distinguishes between learner (LR) and multimedia learning object (LO). For instance, the SQL multimedia lab system replies with a result that includes the result of the execution of the previously submitted solution and additional feedback for the learner.

Interaction is actually central in two dimensions, of which one only has been looked at so far:

- Interaction between learner and learning object is based on multimedia interactions. A simple notation for the abstract interaction specification was illustrated, which can serve as a basis for the media implementation. Here platform and media-type specifics need to be applied. For instance, the Web provides the required submission and reply mechanism in form of HTML forms and HTTP protocol operations.
- Interaction between learning objects, both intra-learning object and inter-learning object sequencing, is based on hypermedia navigation. The learner interacts with the system by navigation through the multimedia learning objects based on an educationally sound navigation infrastructure defined by the instructor.

The second category is usually supported by platform-specific languages and implemented in form of HTML links. Although for instance SCORM sequencing (ADL, 2004), one of the most widely accepted standard for learning technology, is a language to define instructional sequences, platform-specific notations are ignored here, as these are not specific to activity-based learning and training.

Case Study Evaluation

Case Study Deployment

The case study system is actually in use since 1999 to support a database course in an undergraduate computing degree and has been extended since then substantially. The IDLE system has been continuously evaluated since 1999. More than 1000 students from computing, mathematics, and engineering degree programs have used the system so far. IDLE is an e-learning system supporting an on-campus course. In order to understand whether the proposed development framework actually results in an effective e-learning solution can only be decided by carrying out a thorough evaluation of the developed system as the product and also the development process. While software and multimedia aspects such as maintainability and cost-effectiveness are important (and will be discussed), the educational perspective focusing on learner behavior, effectiveness, and the learning experience, which are ultimately the decisive criteria, shall be looked at first.

Evaluation Approach

Whether learners actually use an e-learning and training environment as expected by the instructor is often a question that is difficult to answer. Learner behavior in tool-mediated environments is determined by the learners' motivation, their acceptance of the pedagogical approach and the technical environment, their learning organization, and their activities in the environment. This evaluation should extract factors that make learning successful for the learner and behavior accountable and predictable for the instructor. Success shall be defined in terms of the learner's acceptance (reflected directly though survey data and indirectly through usage data) and effectiveness in terms of high levels of attainment. Meta-level indicators such as motivation and acceptance and analyses based on interaction patterns such as learning organization and usage are distinguished as the main aspects of the evaluation.

The behavior of learners in e-learning environments is influenced by the meta-level indicators 'motivation to use the system' and 'acceptance of the approach'.

- Motivation, the reason to do something, causes the learner to act in some planned and organized way, giving the activities a purpose. It explains when and what purpose learning and training resources are used for.
- Acceptance, i.e. to follow the learning approach and use the system willingly, is crucial for the introduction of new educational technology. It can be explained in terms of surveys and observations about frequency and regularity of usage.

These two factors are usually not the decisive ones, but they provide valuable insights and allow the interpretation of other observations.

Two aspects of the learners' interaction behavior within the learning activity can be distinguished. Firstly, the learning organization addresses the study habits and captures how learners organize their studies over a longer period of time. Secondly, the usage of the system captures single learning activities and embraces how the learner works with the system in a single study session.

- Organization, the way activities are planned and put into logical order, reflects study habits and is guided by the learning purpose. It looks at large-scale interaction patterns.
- Usage, the way the system is used, reflects the actual learning activities of the learner. It looks at small-scale interaction patterns.

Only abstracted patterns of interaction provide an insight into general system usage.

Learner Behavior and Effectiveness

The instruments for the behavior analysis included survey methods to address motivation and acceptance and Web usage mining techniques to capture organization and usage in a Web environment (Pahl, 2004). This combination of methods provides a more complete and accurate picture than uses of survey and learner observation or learner tracking features available in various teaching and learning platforms.

- Motivation. There is a very clear preference for coursework preparation as the main motivation. A Web log analysis shows that the SQL lab feature is mainly used to support coursework and to a lesser extent for exam preparation, which confirms the survey results. More insight into the motivation of the learner's study organization comes from a question regarding the main values of a virtual system; 'always available' and 'self-paced learning' are seen as the key advantages.
- Acceptance. Learners have indicated an overall acceptance of tool-mediated active learning as the pedagogical approach. Comparing traditional and virtual tutorials gives a more differentiated view on acceptance. No favorite emerges; this clearly demonstrates that learners accept virtual tutorials as equally suitable and effective. Another indicator for the acceptance of the approach of self-directed active learning is reflected by frequent and regular usage of the system. While usage mining shows that the tutorial system has not been used frequently and regularly over the whole term, it has however been used intensively in certain periods to fulfill a particular purpose.
- Organization. The study organization (the self-paced learning aspect) shows a just-in-time learning approach of learners with high usage immediately before coursework deadlines and examinations. Web usage mining can give us a clearer picture about the organization. Session classification allows us to determine the purpose of learning sessions, for instance, attending virtual lectures or practicing in virtual tutorials, and to compare the session purposes of different periods. Interactive services are heavily used during a term, but less so for the exam preparation.
- Usage. Besides the long-term study organization, analyzing learning activities within a study session is crucial in order to understand how learners learn. An abstract picture of the purpose(s) of each session can be provided by a session classification technique, but a detailed look at how learners interact with the system, whether they repeat units, or whether they combine interactive elements with lectures is also necessary. Extracting these interaction patterns is important for the instructional designer to compare actual and ex-

pected behavior. The interaction design and its implementation need to facilitate the learner's preferred learning behavior.

Learning behavior analysis is an essential instrument for effectiveness evaluations of tool-mediated active learning. However, effectiveness also comprises aspects such as learner attainment. Exam and coursework results, in comparison with a traditional delivery, can prove the effectiveness with respect to attainment, which is one of the critical success factors. In the system, the learner attainment level with the traditional delivery has been reached since the IDLE system was introduced in 1999. A minimal correlation exists between usage and attainment for the SQL tutorial and lab system, which can be interpreted as an indication for the benefit of using the interactive tutorial and lab features for exam preparation and coursework. In contrast to virtual lecture attendance, where books can serve as an alternative, interactive tools to acquire programming skills are more difficult to replace and, therefore, their usage is beneficial. The high number of students actually using the non-compulsory IDLE system proves this point. Less than 10% of each class decides not to use the system.

Integrated Evaluation and Design Process

An iterative development approach based on formative evaluations that are fed back into design and implementation is necessary to adjust learning processes to the new active media environment. A number of examples shall illustrate this point:

- Activity characterizations in terms of purpose and involvement can be looked at using the behavior analyses (frequency and regularity) and survey results to determine the learners' acceptance of a specific activity type.
- The adequacy of learning goals based on underlying knowledge aspects can be established based on attainment evaluation, but also by classifying learning sessions based on potential learner objectives.
- The specification of the learner-content interaction (which is expected and supported behavior) can be validated using an interaction pattern analysis.
- The suitability of multimedia and channel types can be verified by looking at broader frequency and regularity patterns.

The evaluations help us to scrutinize essential design and implementation decisions. Evaluations also address unexpected aspects, such as just-in-time learning patterns, that would have to be readdressed in revisions of the design.

Software Development Issues

Usability is an important software quality. The literature distinguishes two aspects of usability – technical and pedagogical usability. The interaction infrastructure is the technical side that addresses technical usability, whereas the interaction space based on the interaction model needs to be based on pedagogical usability aspects. Technical aspects such as media type and interaction designs are fitted into a pedagogical framework of learning activity types in order to allow the interaction infrastructure to support learning interactions defined in the interaction space. Classical Web usability issues, such as page layout, are neglected here, allowing more emphasis to be put on pedagogical issues such as motivation, acceptance, organization, and usage.

This chapter has suggested some development techniques for e-learning systems. Although the focus here was on the implementation of active learning, addressing software technologies aspects should not be neglected. One central lesson learned, supported by other authors such as Palmer and Tulloch (2001), concerns software properties. Often e-learning systems are research prototypes and showpieces. These systems have turned out to be costly and difficult to maintain, if used on a day-to-day basis. The development of easy-to-use infrastructures and mainstream system designs should be favored if the maintainability of learning objects and e-learning applications and not the exploration of new technologies is the objective. Even in the latter case, the technologies might remain in service for several years and, consequently, need to be maintained as well. These observations should always be considered in the design of learning objects when choices regarding media types or infrastructure technologies are made.

Discussion

Although the overall experience with the case study system is positive, some problems have occurred. Therefore, important limitations and weaknesses for learners, instructors, and developers shall be discussed.

- The system replaces the direct interaction between learner and instructor to some extent. The level and quality of feedback for the learner is consequently reduced, although it is worth noting that the learner can get more feedback at a lower level due to the constant availability of the system. The instructor equally receives less direct feedback and has more difficulty in monitoring the course activity.
- The analysis of usage patterns and learning organization shows a just-in-time learning behavior. Although instructors prefer a more regular approach to learning, this reflects common learner behavior. The motivation is a central factor in what determines a successful learning experience.
- The cost of developing and deploying an e-learning system such as the presented one is often prohibitive for a single organization or provider. Only sharing and the reuse can provide a solution in many cases.

These more practical limitations are completed by a more foundational limitation that was already discussed in the literature review. Learning is not an activity and cannot be supported directly. The system discussed here can provide virtual versions of tools and their environment in some contexts well, but the instructor is more difficult to replace. Only some basic instructor tasks have been attempted in order to relieve the instructor from these, such as feedback and personalized guidance in a range of standard situations.

A central question that needs to be discussed is the transferability of the approach to other topics and other groups of learners. In the context of active learning and training support, similar technologies have been used for other subjects beyond the computing domain. In engineering and sciences, so-called virtual instruments are often used to enhance training aspects. Learners can, for example, carry out mechanics or physics experiments and can analyze the resulting data using these virtual instruments, which usually combine a multimedia representation with an analytical tool. Another example of e-learning systems for active learning and training are simulators used to train people in the usage and handling of complex systems ranging from aircraft to medical devices. These combine multimedia content with real-time computation features. The pedagogy does not necessarily need to be changed, as the principle is still a learning experience based on interactive exercises and problems. Only the degree of personalization might vary from application to application.

The presented technology is transferable to other groups of students. The case study has been used by students across a number of backgrounds, though predominantly from computing. With increasing computer literacy of younger learners and even advanced IT skills of third-level students in technical disciplines, these media- and interaction-intensive e-learning systems are common. The case study system is accessible using Web browser technology and does not exceed the complexity of standard PC desktop tools, making it in principle accessible for a broad range of learners.

Conclusions

Essential problems that a content developer faces in the development of learning objects for active learning and training relate to very practical aspects. What media types work for a particular learning activity or at what cost can these learning objects be developed, deployed, and maintained are central design issues. A development approach (See Figure 1) for active learning objects has been presented that provides a framework to answer some of these questions and that highlights the difficulties specific to active learning and its implementation through interactive educational multimedia.

Experience shows the importance of knowledge-or skills-level interaction, i.e. interaction that is meaningful within the context of the subject domain (Ravenscroft et al., 1998). The interaction model that has been presented is an important feature that captures central elements of a good interaction design by linking learning interaction (the educational perspective, i.e. the interaction space) with multimedia interaction (the technical perspective – the interaction infrastructure). In addition to the interaction model that structures an interaction space, an interaction infrastructure is needed. The interaction infrastructure enables learners to achieve their learning objectives in the system. The interaction infrastructure consists of media to convey content, a navigation topology for sequencing, and learner-media interactions for knowledge- and skills-level activities. The evaluation of the approach has looked at meta-level indicators and usage observations. Both are important to decide whether the learning objectives can be realized and whether the learning object implementation is effective and successful.

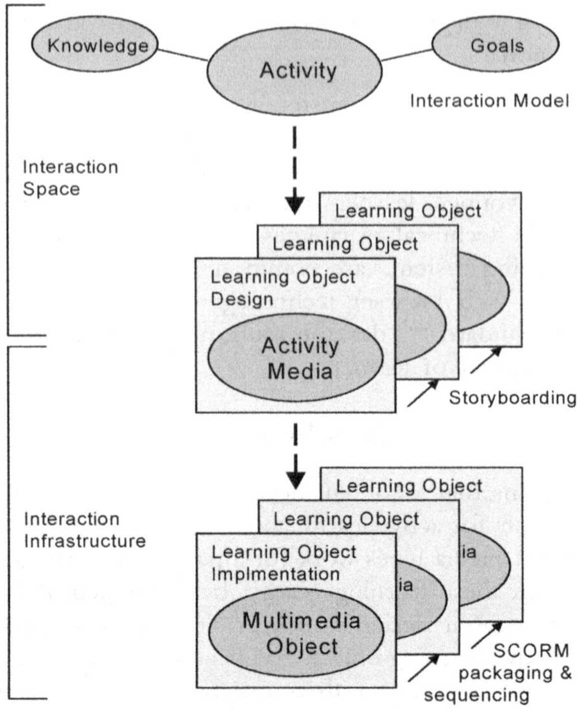

Figure 1. Active Learning Object Development Process.

The development approach influences both the educational quality and the software quality of an e-learning system. The method is feasible, allowing the implementation of an effective e-learning solution. The evaluation, however, has also highlighted critical aspects:

- On the educational side, learning behavior is difficult to predict and model. Only an iterative approach involving design, implementation, and formative evaluation provides a solution. Instructional design is difficult in novel, technology-supported environments. Designing instruction such that the learner's behavior corresponds to the instructor's expectations is a challenge.
- On the technical side, the costs of maintenance and change are often underestimated, resulting in these aspects being ignored in learning object design. Change should be considered in the design from the outset. This is in particular the case, if advanced techno-

logical solutions based on interactive educational multimedia are involved.

Some observations show the link between interactive educational multimedia design and our formative evaluation results. The more interactive the features, the more they are used. The SQL feature is a good example of this, which supports the hypothesis of active learning as an effective form of instruction. The media types and navigation infrastructure to support interactivity not only on a technical level, but also in an educationally sound an effective way, need to be scrutinized through evaluation as our experience with initially unexpected and undesired behavior suggests.

One of the lessons learned from the case study is that a development and management methodology for e-learning technology is needed. The objective here has been to raise an awareness of the solutions, but also the technical problems in the development of e-learning solutions. Although the discussion has focused on a computing subject, the approach is not specific to computing, and can be applied to any other domain. The media types addressed are standard media supported by Web infrastructure. What might vary from subject to subject is the preferred media type, whether it is text-based if more formal languages are involved or whether more graphical types prevail. Overall, this chapter has pointed out some aspects of active learning object development, see Fig. 1. A full design methodology would have been beyond the scope of this investigation.

References

ADL. (2004). *Sharable Content Object Reference Model* (SCORM) – Content Aggregation Model (CAM), Run-Time Environment (RTE), Sequencing and Navigation (SN). http://www.adlnet.org/index.cfm?fuseaction=scormabt. (visited 06/04/2006).

Boyle, T. (1997). *Design for Multimedia Learning*, Prentice Hall, Europe.

Boyle, T. (2003). Design principles for authoring dynamic, reusable learning objects. *Australian Journal of Educational Technology*, 19(1), 46-58.

Elsom-Cook, M. (2001). *Principles of Interactive Multimedia*. London: McGraw-Hill.

Fowler, C.J.H., & Mayes, J.T. (1999). Learning relationships: from theory to design. *Association for Learning Technology Journal*, 7(3):6-16.

Heller, R.A., Martin, C.D., Haneef, N. & Gievska-Krliu, S. (2001). Using a Theoretical Multimedia Taxonomy Framework. *Journal of Educational Resources in Computing*, 1(1).

Jonassen, D.H. & Mandl, H. (Eds). (1990). *Designing Hypermedia for Learning*. Berlin: Springer-Verlag.

McLoughlin, C., Winnips, J. C., and Oliver R., (2000). Supporting Constructivist Learning through Learner Support On-line. In *Proceedings of the World Conference on Educational Multimedia, Hypermedia and Telecommunications EDMEDIA 2000*, pp 674-680.

Melis, E., Weber, M. & Andres, E. (2003). Lessons for (pedagogic) usability of elearning systems. *Proceedings Conference on E-Learning in Corporate, Government, Healthcare, and Higher Education eLearn-2003*, pp 281-284.

Moore, M.G. (1992). Three types of interaction. *The American Journal of Distance Education*, 3(2), 1-6.

Murray, S., Ryan, J. & Pahl, C. (2003). A tool-mediated cognitive apprenticeship approach for a computer engineering course. *IEEE International Conference on Advanced Learning Technologies ICALT'03*, 2-6.

Nardi, B. (ed.) (1997). *Educational Context and Consciousness: Activity Theory and Human-Computer Interaction*. MIT Press.

Northrup, P. (2001). A framework for designing interactivity into Web-based instruction. *Educational Technology*, 41(2), 31-39.

Ohl, T.M. (2001). An interaction-centric learning model. *Journal of Educational Multimedia and Hypermedia*, 10(4), 311-332.

Okamoto, T., Cristea, A. & Kayama, M. (2001). Future integrated learning environments with multimedia. *Journal of Computer Assisted Learning*, 17, 4-12.

Pahl, C. (2004). Data Mining Technology for the Evaluation of Learning Content Interaction. *International Journal on E-Learning*, 3(4), 48-59.

Ravenscroft, A., Tait, K., and Hughes, I. (1998). Beyond the Media: Knowledge Level Interaction and Guided Integration for CBL Systems. *Computers & Education*, 30(1/2), 49-56.

Sims, R. (1997). Interactive learning as "emerging" technology: a reassessment of interactive and instructional design strategies. *Australian Journal of Educational Technology*, 13(1), 68-84.

Trikic, A. (2001). Evolving open learning environments using hypermedia technology. *Journal of Computer Assisted Learning*, 17, 186-199.

Virvou, M. & Tsiriga, V. (2001). An Object-Oriented Software Life Cycle of an Intelligent Tutoring System. *Journal of Computer Assisted Learning*, 17:200-205

Author

Claus Pahl is a Senior Lecturer and the leader of the Web and Software Engineering research group at Dublin City University, which focuses on Web technologies and e-learning applications in particular. Claus has published more than 120 papers including a wide range of journal articles, book chapters, and conference contributions on e-learning. He is on the editorial board of the International Journal on E-Learning and is a regular reviewer for journals and conferences. He has extensive experience in educational technologies, both as an instructor using technology-supported teaching and learning at undergraduate and postgraduate level and as a researcher in Web-based learning technology.

Buzzetto-More, N. A. (Ed.). (2007). *Advanced Principles of Effective e-Learning*
Santa Rosa, California: Informing Science Press. (pp. 161-183)

Chapter 6

Managing e-Learning Quality in Practice

Luca Botturi, Chiara Succi, Lorenzo Cantoni and Alessandro Inversioni

Introduction

Hamlet's doubt about "to use or not to use" information and communication technologies (ICT) in education is gone. That was no easy work, and it wasn't made easier by emphatic marketing-oriented messages or by the promise that technologies would be the magic wand to make higher education better. Once it was clear that technologies – unlike king Midas – do not turn all they touch into Harvard or Stanford and that they do not curse their users, all institutions have begun to use them at different paces and following different paths. And as it turned out to be, the real challenge is integration.

The emergence of quality as key issue is a clear indicator of real growth (Cantoni & Succi, 2003): from Hamlet's doubt to the first trials, and from that on to a continuing process of integration into the fabric of academic life. In the last years, we witnessed a rapid development of e-Learning activities in European universities, as well as the launch of national support programs in most countries (Bates, 2001; Van der Wende & van der Ven, 2003). There is a large number of studies on the evolution of the higher education system in Europe and worldwide (cf. De Boer et al., 2002) and most of them try to identify the role of e-Learning, its opportunities for development and main trends (Coimbra Group, 2002; Collis & Van der Wende, 2002).

While at first the attention of researchers was captured by the issue of effectiveness of the use of ICT, by comparing "traditional" courses with courses using them (Phipps & Merisotis, 1999), more recently it is becoming clear that the real point is not whether e-Learning is effective, but whether it is efficient, and under what conditions. Once universities focused on the integration issue, a different trend emerged. There is no unique recipe for e-Learning implementation at all sites: each time it is necessary to assess the opportunities and weakness of new technologies in that peculiar educational context. Consequently, a crucial and more promising research area seems to be that concerned with teaching and learning quality. The problem of using e-Learning or not is an old one – what is at stake for those working in e-Learning is how to implement it in order to offer a higher quality learning experience (Phipps & Merisotis, 2000).

However, quality control is a cost, and might be a huge one. Most of all, such cost can be perceived as *external* to the project – i.e., it does not bring and product or direct benefit to teachers – and its size can hinder its application.

This chapter presents a lightweight method for managing quality throughout the e-Learning design and development process in a sustainable way. While the method emerged from the practice of a small but busy e-Learning support unit in the academic setting, its main features are generalizable also to different settings. The next section provides a synthetic definition of quality, useful in order to understand the specific feature of the method presented afterwards. Section 3 reports the results of a benchmark study, in order to assess the current situation of e-Learning quality evaluation in the academic setting. The method is presented in section 4 and is followed by a case study in section 5. Finally, section 6 compares the method with other quality evaluation systems.

Quality Definitions, Perspectives and Strategies

Quality issue is not a new issue in the academic agenda. Since the foundation of the first universities in the Middle Ages in Europe, their mission has been understood as to offer high quality teaching and learning experiences (Vroeijenstijn, 1995). e-Learning has just pushed the discussion forward. According to Barnett (1992), the debate depends on the modern *quality gap*: the increasing number of students, the widening offer of curricula and new education opportunities (due also to ICT) carried out with unvaried governmental funds, have pushed institutions to reflect on the quality of their learning and teaching offer. The need of a systematization of assessment and evaluation processes has emerged as a consequence (Maassen, 1997).

Quality is a rather fuzzy concept. Ehlers (2004) proposes to articulate a definition on three dimensions, sketched in Figure 1: (a) different meanings of quality; (b) different actors' perspectives; and (c) different levels of quality.

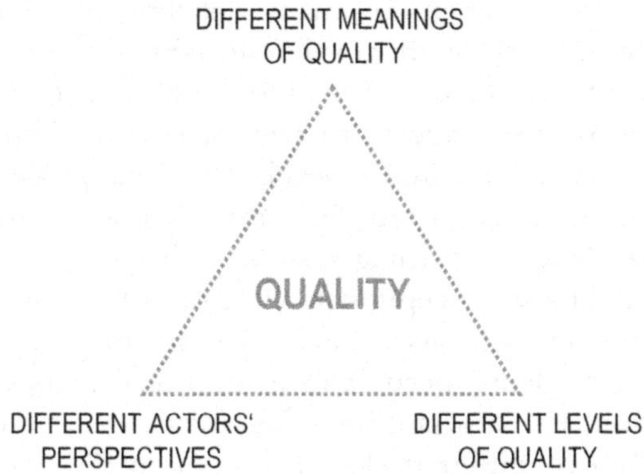

Figure 1 - Multiple perspectives on quality in eLearning (adapted from Ehlers 2004)

Meanings

The meaning of quality is related to the context and to the object that should be "of high quality". Exceptionality, perfection, the achievement of goals, the cost/benefit ratio or change are all possible facets of a cluster of meanings. For a service, such as a travel agency, quality will be examined based on customers' expectations: high quality will be that of an exceptional service that no other provider can guarantee. On the other hand, when it comes to the quality of a product, such as a pair of shoes, quality is the level of perfection, intended as lack of defects. In education, as in many other fields, quality is the achievement of predetermined (learning) goals; in other domains, such as in the public administration, where the money belongs to citizens, quality is measured considering the cost/benefit ratio. Finally, interventions, such as a chirurgical operation, need to result in change and have a positive impact in order to be judged successful.

In education, goal achievement is often the core idea behind both formative and summative evaluations (Morrison, Ross & Kemp, 2003). But when it comes to e-Learning, other items, such as Learning objects or digital learning materials, require a product-oriented perspective – this is indeed one of the issues tackled in the method presented below.

Stakeholders

Moving to the second dimension, many stakeholders are involved in a learning activity and each of them is interested in different aspects (and also uses different meanings). A nation funding a learning program through its government expects to have a positive impact onto the society (change). At the institution level, the whole process of a learning activity is considered, from the effectiveness of the enrolment phase to the tracking of learning results, as having an impact on the institution's reputation (exceptional service). On a narrower scale, teachers and instructors try to achieve teaching goals with cost/benefit logic combining students' needs and institutional requests, while student's satisfaction, an important but often over-considered indicator of quality, is a blend of different elements: content meeting expectation, interest, time management, trust, etc. Finally, also parents are very important stakeholders: they are, often, the main sponsor and promoter of students' education.

Levels

Different levels of quality in education are usually referred to Kirkpatrick's Taxonomy (1994). He distinguished four levels of quality that were integrated by Phillips (1996):

1. Reactions: learners' reactions need to be taken into consideration in an evaluation process as a first feedback to the instruction.
2. Learning: learning results can be tracked through different modalities, such as tests, interviews, monitoring, etc.
3. Transfer: in many contexts, such as vocational training or lifelong learning, the point is to make a difference on the working field, transferring new knowledge in the everyday practice.
4. Results or ROI: did the costs really bring what was expected? Especially in the corporate sector, the final outcome of a learning initiative should be evaluated also measuring the Return On Investment in economical terms.

Quality Benchmarks and e-Learning Management

The intrinsic complexity of meanings, stakeholders and levels connected to quality gave rise to a wide range of measurement strategies, used by private and public organizations (Van Damme, 2002). Studies and researches have produced set of standards, benchmarks or guidelines in order to lead the evaluation of e-Learning activities (Massy, 2002; ASTD, 2004).

In particular, a research on e-Learning quality within universities was conducted with a mandate of the Swiss Virtual Campus Programme (Cantoni & Succi, 2003; Cantoni, Lepori & Succi, 2003). The study involved the Swiss universities and those in Baden-Württemberg (Germany), Lombardia (Italy), Rhône-Alpes (France) and Catalunya (Spain), with a total of 52 Universities.

The research activities exploited the 24 benchmarks proposed by the IHEP in its study *Quality On The Line* (QOL; Phipps & Merisotis, 2000), commissioned by the US National Education Association (NEA) and Blackboard Inc.

Data suggest that e-Learning experiences started in an institution- and teacher-centered frame, and are yet to move towards a more learner-centered approach. In particular, a major effort is to be devoted in designing and implementing evaluation and assessment tools and processes; despite its already decade-long history, e-Learning is new to many institutions, which focus more on the creation of a suitable environment than on its evaluation. In any case, as long as this parameter remains neglected, it will be quite difficult to judge the cost/benefit ratio of e-Learning initiatives.

This last consideration concerns also the e-Learning management issue: as long as e-Learning is offered through single and isolated initiatives, the issue of its medium-term economic and organizational sustainability can't be properly tackled. The same aspect is stressed also by the wide dispersion of answers concerning the "whys" of e-Learning implementations: no answer was shared by more than 33% of respondents, and those related to costs – hence specifically managerial – scored poorly. Also incentives for professors who decide to enter the "terra incognita" of e-Learning have to be better defined, as well as procedures for accreditation and copyright issues.

Results were largely confirmed by a further research extended to all the 77 Italian universities and to European universities of applied sciences (Lepori & Succi, 2004; Cantoni & Esposito, 2004).

A Sustainable Approach: the eLab Approach

In this context, the eLab (n.d.) strived to move one step forward and develop a lightweight method for making e-Learning quality evaluation a doable and sustainable daily practice. The e-Learning Lab is a joint service of the University of Lugano (USI, Università della Svizzera italiana) and the University of Applied Sciences of Southern Switzerland (SUPSI, Università Professionale della Svizzera Italiana). Currently, the eLab supports 18 financed inter-university e-Learning development projects, runs a LMS with about 1000 courses and over 3000 users, offers training to the teaching staff of USI and SUPSI, and carries on open source e-Learning application development.

The daily needs of the eLab led to the idea of evaluating the quality of e-Learning courses considering its different but intertwined meanings, and in a time- and cost-effective way. The result is a method that focuses on (a) quality of e-Learning tools and learning materials as a "perfect product"; (b) quality of an "exceptional" e-Learning support service; and (c) quality as the achievement of learning goals with a reasonable cost/benefit ratio. In practice, the eLab methodology is based upon three main pillars: (a) technical quality, i.e., the usability of digital learning materials and of the online environment; (b) integration, i.e., to what extent and how e-Learning activities are deployed within the framework of a single course, completely online or blended learning; (c) summative evaluation, i.e., the *ex-post* evaluation of a course at the end of its lifecycle.

In order to properly understand the method in its practical implications, it is necessary to have a glimpse of how the activities are organized at the eLab. This is also paramount in order to understand the case study presented in the next section. However, while the method was specifically developed at the eLab, it is both domain-independent (i.e., it can be carried out for courses on different subject matters) and LMS-independent (i.e., the evaluations can be carried out in different LMS environments such as WebCT Vista, Moodle, etc.).

Evaluation and Fast Prototyping

E-Learning design and development teams at eLab are usually composed of one instructional designer, a web programmer and a graphic designer, that work closely together with one or more subject-matter expert. Each member of the staff is normally engaged at the same time in three or four projects – the lack of time is constant, so that even if the project plan entails some time for quality evaluation, often this time is devoted to other issues, such as bug fixing. The lack of resources, mainly time, is what made quality a sort of *chimera* at eLab, as in many other similar units. The way out of such a critical situation was the development of a simple conceptual framework implemented by lightweight procedures that make quality evaluation a benefit: the investment in terms of time is rewarded by useful feedback for more successful projects.

eLab teams use the fast prototyping model (Botturi et Al., 2006). Thanks to this model, after the requirements are defined, and an instructional strategy is developed, the implementation starts in quick cycles in order to let stakeholders discuss about real instructional products. In this process, quality evaluation is a crucial step of the cycle, both for measuring the quality of the technical product (e.g., a CD-rom) and its impact during tests with real students.

Steps, Goals and Instruments

The methodology is composed by three different steps: prototype evaluation, process evaluation, and summative evaluation, as sketched in Figure 2. The three steps are closely related but have different goals and exploit different instruments. They also refer to different stages of the design and development process, as shown above: the prototype evaluation refers to the product cycle, while the process evaluation refers to the actual integration of e-Learning materials and methodologies in a real course, and pertains to the process cycle. The last step, summative evaluation, takes place at the end of the project.

The following paragraphs provide the details for the three steps.

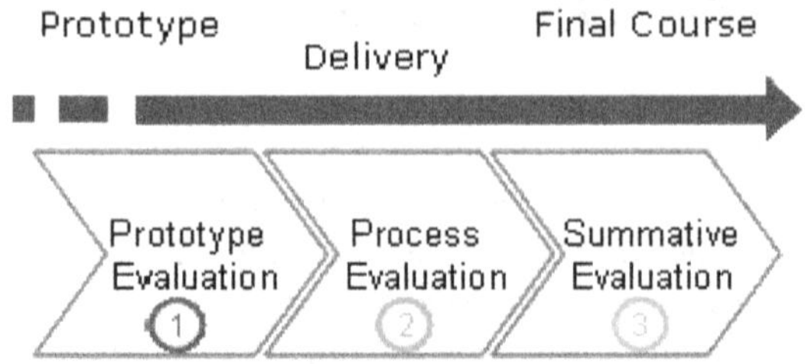

Figure 2 - Overview of the evaluation method

Prototype Evaluation: Technical Quality

The goal of this step is providing feedback for improving the technical and usability features of all digital resources being developed in the project. The prototype evaluation starts when (a) the prototype is finalized and (b) project partners and team members agreed to have formal usability evaluation, i.e., they are ready to get a structured feedback and revise their plans.

The prototype evaluation consists of two activities:

1. In the *Technical Inspection* an evaluator goes freely through the application following a technical and usability checklist. Her/his task is to comment wherever there is a usability issue. The checklist was created on MiLE+ heuristic inspection basis (Triacca et Al., 2004), and has been adapted to the e-Learning environment.
2. The *Expert Review*, which follows the technical inspection, is a kind of "synthetic history of use" of the application. The same evaluator goes through the application following the indications of specific scenarios (e.g., a students right before submitting an essay), and tasks (find the dropbox), always looking for usability breakdowns. S/he reports all issues following a set of instruments: a library of scenarios developed by the project's instructional designer, the task library, and the User Experience indicators library (Triacca, Inversini, & Bolchini, 2004).

The output of this phase is a structured usability report, containing a ranked list of errors and issues, along with specific proposal for fixing from the evaluator. The idea is that this is a first, product-oriented feedback from an external expert reviewer, who provides input to the project in a structured way, using usability guidelines and principles, and who is helped to get "in the shoes of the students" by the scenarios prepared by the instructional designer.

Process Evaluation: Learning Quality

The goal of the process evaluation is to provide feedback on the actual practices of students in the learning environment and to assess its effectiveness in terms of satisfaction and learning. Differently from the prototype evaluations, this step does not evaluate a specific learning resource or online environment, but a blended learning course as an instructional activity.

The process evaluation starts once all learning materials are finalized as products, the course is ready, and the project partners and team members agree to undergo the evaluation. In practice, during the course supply a bunch of activity are planned. At the end of the process, partners review feedback and make decisions.

The evaluation starts before the course starts, with the instructional designer and the course instructor (or instructors) sitting together and describing the course scenario: who are the students? How will they use the online resources? What are the instructor's expectations? What is s/he planning to do? Right before the end of the course (usually before the final evaluation, if planned) students are surveyed with a standard online questionnaire built around Kirkpatrick's (1999) first two dimensions: reactions and learning (with some hints as to perceived transfer potential). After the end of the course, the instructional designer interviews the instructor(s) once more: what has happened? What did you do? Did the course meet expectations? All this information is put together with the course grades and a standard analysis of log files in the final evaluation report, which is then discussed with the project team. Figure 3 sketches the activity plan for step 2. All activities are supported by standard documents and guidelines that make easy to carry out the evaluation.

The idea is that the process evaluation can be done more than once, in repeated test phases, and that its results will be incremental.

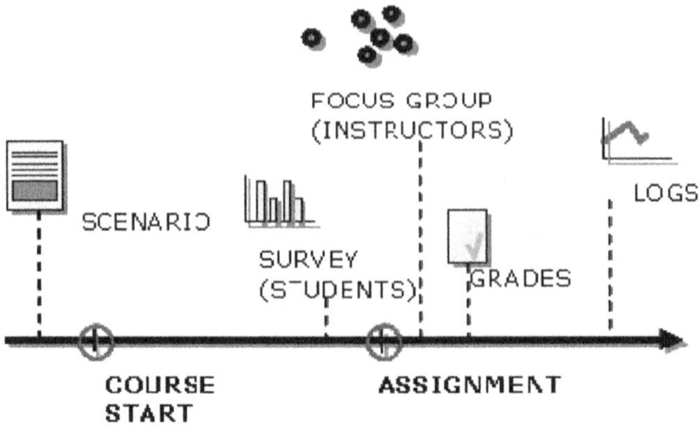

Figure 3 - activity plan during process evaluation

Summative Evaluation

The goal of the summative evaluation is to set a milestone at the end of the project that summarizes all the evaluation activities done so far with a final picture of the overall quality of the project. The results of this step are useful mainly in order to present the outcomes to important stakeholders – such as the financing body, which can be sometimes responsible for a continuation of the project itself.

The idea is to monitor the final test phase of the project through the same activities describe above in the process evaluation. Its results will then be compared with those from previous process evaluations into a final report targeted not to the project team as feedback but to external reviewers.

Costs

As mentioned above, this e-Learning quality evaluation method was developed and tested in a small unit with a lot projects going on at the same time. Time and cost effectiveness were therefore crucial criteria in its development, much more than exhaustiveness or precision: the choice was made for a continuing monitoring of the projects resulting in a set of limited but comparable data, than sporadic more complete but non-comparable observations. In this way, quality can become a useful habit, with techniques that can be improved, refined or expanded when necessary. For example, particularly critical or unclear issues can be further investigated through ad-hoc surveys, interviews, cognitive walkthrough sessions, etc.

This concern has pushed also to define specific roles for each step, and to associate time resources to each of them:

1. The project *instructional designer* is in charge of developing project-specific scenarios and tasks (starting from a generic library) to guide the evaluator in step 1. In step 2 and 3, s/he is responsible of collecting all data (the two interviews, survey, grades, log analysis). S/he is also the author of all reports (which can be revised by an external reviewer for critical cases). The instructional designer is expected to spend 1-3 hours of work on step 1, and 4-5 hours in step 2 and 3.
2. The two activities in the prototype evaluation step are carried out by an *external designer*, i.e., on of the lab's designers who is not directly involved in the project. S/he is expected to invest 3-4 hours in step 1, depending on the dimension of the product and on the number of issues.
3. The involvement of course instructors is paramount in steps 2 and 3: they collaborate in sharing their view on the course before and after it takes place, and also help instructional designer to understand the student's feeling with the project products. Their effort is estimated as 2 hours.

The amount of time required by each role does not include the time required for designers to learn the method. This is however a little investment, as a 1-day seminar has proved enough for getting people started.

Table 1 presents the set of tools used in each step.

Table 1 – Tools for each step

Step	Tool	User	Author
Step 1	Technical heuristics	Evaluator	eLab (from MiLE+)
	Use scenarios and tasks	Evaluator	Instructional designer
	User Experience Indicator (UEI)	Evaluator	eLab
	Generic use scenario and tasks	Instructional Designer	eLab
	Report template	Instructional Designer	eLab
Step 2 and 3	Course Scenario of use	Instructional Designer	Instructor + Instructional Designer
	Student Survey	Instructional Designer	eLab
	Assessments	Instructional Designer	Instructor
	Log data	Instructional Designer	LMS
	Teacher interview guidelines	Instructional Designer	eLab
	Report template	Instructional Designer	eLab

Case Study: Hear and See

The Swiss Virtual Campus (SVC) is a Swiss federal funding program aimed at introducing e-learning in higher education (SVC, n.d.). Hear and See (Hear and See, n.d.) is a SVC project running from September 2005 to December 2007. It is led by the Institute for Media and Journalism of the University of Lugano, in partnership with the Universities of Bern, Fribourg and Lausanne in Switzerland. It also enjoys the collaboration of the University of Applied Sciences of Luzern for graphic and interface design and of the Archives of the Swiss public television system (SSR-SRG) for the reuse of historical audiovisual materials. The technical support for implementation and maintenance is provided by the eLab. The main goal of Hear and See is producing a set of online resources about media history, media analysis and media biographies for students in Media Studies, Communication and Journalism throughout Switzerland. Its added value is (a) providing teachers and students with access to archive and rare audiovisuals materials; (b) offering interactive hand-on exercise on different topic related to media analysis; and (c) supporting online communication among students and teachers.

Hear and See developed a number of resources, including 20 online units presenting theory and exercises on different topics. These units are connected with a streaming media server that supports the download of audiovisual materials. The units are packaged as Learning Objects (Botturi, 2005), delivered to students via a LMS. The online units are accompanied by a set of web resources, namely an online glossary, a blog and an online media biography application.

Hear and See followed the eLab fast prototyping approach (Botturi et al., 2006). A first kick-off meeting presented the requirements analysis and the scenario developed by the project leaders, and 4 months later a second meeting inspected the first prototype of one of the online units. By month 6 the first online unit was ready along with the first interactive lab, which were tested a couple of months later, when the first class of students was available. This process was then repeated at a faster pace for all online units and interactive labs. The other resources (the glossary, the media biography application, and the blog) were also prototyped at different stages of the project.

Critical Quality Issues

The quite wide array of resources developed for Hear and See was conceived in order to accommodate different teaching and learning needs, as emerging at partner institutions. Considering such differences is clearly a key challenge for quality measurement. Also, Hear and See aimed at producing flexible and usable resources. Namely resources that (a) do not require special hardware, software or Internet connection; (b) have easy interfaces for students of different languages; and (c) are usable in different learning contexts and scenarios. Controlling their quality *during* the design and development process, and not afterwards, was a key issue for a successful development.

Quality evaluation in the Hear and See project followed the pattern of the model presented in this chapter and its three steps. For simplicity's sake, from now on we will focus on the evaluation of online units, understanding that the evaluation of the other items developed in the project followed the same process.

Step 1

As soon as the first prototype unit was developed, it underwent a complete usability evaluation. In this case it was possible to run two heuristic-driven evaluations and two scenario-based inspections, whose results identified a list of usability errors or issues, organized along the method's main heuristics. Some of them were local problems, for example, the lack of indication of authorship on some pages; others were general problems, for example the difficult interpretation of some of the interface's icons. The project's instructional designer collected all issues into a final list, indicating possible solutions for each of them. The issues were also prioritised and grouped according to the amount of work required in order to fix each of them. While some issues required minor changes (e.g., adding a caption to a figure), other required a systematic review of the whole unit (e.g., adding a meaningful ALT tag to all graphic elements), or even an additional work with the graphic designer (e.g., producing a new icon, or defining a new match between different types of learning materials and the color codes used to identify them).

The list was then discussed with the project team: basically, all issues that allowed a quick but effective fix were solved, and additional effort was put in some of the other issues, deemed highly relevant for students' use in the foreseen scenario. Step 1 took place before any tryout with real students, so that when the first class put their hands on the prototype online unit as part of a real course, all major usability issues had already been fixed.

Step 2

The first real use case was a class of 40 students in Communication, attending the TV Programming course in their 3rd year at the University of Lugano. The instructional designer followed the method's indications, and collected all necessary data and documentation: (a) the scenario description; (b) data from the student survey at the end of the course; (c) data from the LMS tracking system; and (d) the concluding interview with the instructor. Figure 4 reports the activity flow diagram in the scenario description (the visualization used is taken from E2ML – Botturi, 2006).

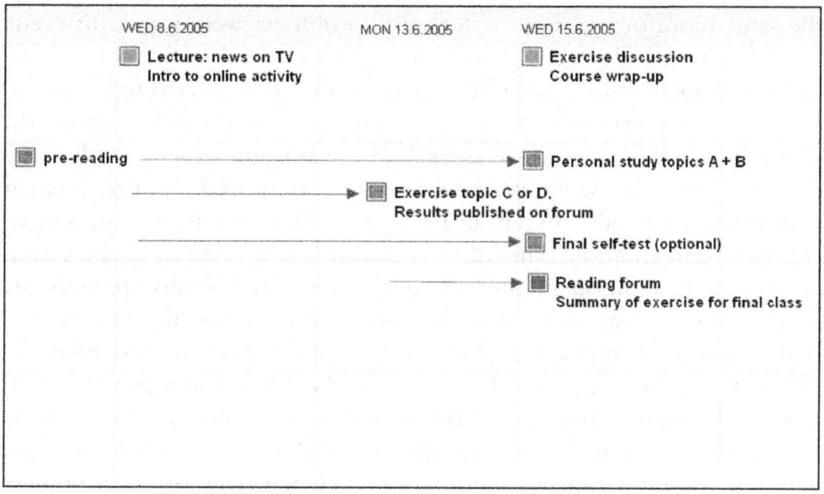

Figure 4 – Activity flow diagram

The first point emerging from the test was that the students encountered no technical problem, both when working from home and from the university campus. Although the online unit actually offered more materials than those required, the general view emerging from the data allowed determining that students strictly followed the pattern indicated by the teacher with rare exceptions. The average time students spent on the online unit (5h) matched with the teacher's expectations, and they were actually interested in the topic and enjoyed working online.

These results were then discussed with the course teacher with two different objectives: (a) identifying improvements for Hear and See courseware, e.g., providing printable versions of long texts; and (b) reflecting on the integration of the online units in the course structure, identifying barriers and new possibilities. The online unit was revised accordingly to the design guidelines emerged in the test, which were also considered for the development of the other online units.

Three months after, a set of four online units, including the one already tested in Lugano, were used at the University of Fribourg, in a course of Media History for Journalists. The scenario here was completely different: only 12 students had the possibility to access the online units as optional material for in-depth study of some course topics. The evaluation followed the same procedure, and the outcomes, except for the same (comforting) lack of technical problems, were rather different. First, students only used the first module presented in the course, and did not actually used the other ones. It was then discovered this was due to a naïve expectation of the teacher's, who just talked about the online modules during her classes, and did not implement a follow up session. Secondly, while students in Lugano enjoyed the presence of audiovisual materials, student in Fribourg were happy to find structured course documentation. This difference can be traced back to the different teaching tradition in the two universities: in Lugano students are used to receive handouts from teachers, which is not the case in Fribourg, where seminars are more discussion-oriented. In this case, the different context – and not the quality of the e-learning materials or the type of learning activity – produced different results. This interpretation of the student's survey results emerged from the final interview with the teacher; otherwise the difference between the two student groups would have been inexplicable.

While Step 1 was run on the first prototype, and allowed adjusting the online unit template, within the context of Hear and See Step 2 was also run in different courses using different items developed in the project. This allowed (a) creating a base of data about the uses, issues and outcome of the project; and (b) identifying recurring issues that were put under strict observation (e.g., through interviews) and, when possible, fixed.

Step 3

Hear and See is now halfway through its lifetime: in December 2007 it will reach the end of the financing period and possibly find other ways to maintaining and further develop its digital resources. The possibilities it has are twofold: find a further financing from the original program, and/or finding a sponsor or support from a private, enterprise or institution. Independently from the results, in both cases the base of evaluation results it collected during its first two years provides a solid advantage: on the one hand it demonstrates the project's willingness to achieve sound results and to continuously improve; on the other, it allows presenting the project's strengths with real data at hand. Both are possible thanks to the availability of a lightweight method that allows continuing evaluation at reasonable costs.

About Costs

As we mentioned above, the practical problem with evaluating e-Learning is costs, where evaluation is time consuming and requires experts. This results in confining the evaluation to the end of the project or to specific phases. The lightweight method presented here allows making the evaluation process continuous offering a good trade-off between costs and results. Also, following a standardized process, supported by simple tools allows sharing part of the job with non-experts.

For Hear and See, Step 1 required 4 days for the double evaluation and the inspection, and half a day for compiling the usability report. For each evaluation in Step 2, we calculate 1 day of work of the instructional designer.

Certify and Support: A Comparison

The eLab evaluation method is process-based. It can be compared with a wide range of measurement strategies developed by international private and public organizations, such as Quality Assurance, Quality Audit, Quality Assessment, Accreditation or Benchmarking (Van Damme, 2002). Studies and researches have produced set of standards, benchmarks or guidelines in order to lead the evaluation of e-Learning activities (Massy, 2002). The European Commission supported four projects within the e-Learning Action Plan (EQO, SEEL, Qual E-learning, SEEQUEL), while other associations as ADEC, AFT and ASCLITE, ASTD, IHEP, EFQUEL, and EFMD promoted the development of a set of parameters to measure quality in e-Learning.

In particular the ASTD (American Society for Training & Development) method proposes a set of standards for the E-Learning Courseware Certification (ECC). Rather than certifying a particular technology, the ECC intends to certify the quality of the learning experience. It is focused on the ability of a course to teach – it does not certify content, rather it looks at the relationship between users, technology, and content. In fact, ASTD declares it would be nearly impossible to certify that all of the content contained in a course is true and accurate. However, standards can be used to determine if content exists for all course objectives, if the content is well presented, if the course reinforces learning with sound instructional design practices, if it uses predictable technology, and if it supplies an engaging learning environment.

ASTD standards are grouped into four principal categories to reflect the various elements of courseware design.

1. Interface Standards address the relationship between the learner and the courseware itself. There are five interface standards.
2. Compatibility Standards address the relationship between the courseware, the operating system, and related applications. There are four compatibility standards.
3. Production Quality Standards examine the quality of the courseware's text, graphics, grammar, and visual presentation. There are two production quality standards.
4. Instructional Design Standards examine the relationship between the course purpose, objectives, instructional content, instructional methods, and the learner. There are seven instructional design standards.

Another example is the effort made by ISO/IEC (2005) to harmonize the various approaches used around the world for assessing the quality of e-learning initiatives. The standard harmonizes the international conception of e-learning quality by creating a coherent inventory of the diverse processes which affect the attainment and preservation of e-learning quality. These processes embrace all e-learning application scenarios, such as content and tool creation, service provision, learning and education, monitoring and evaluation, and lifecycle stages – from continuous needs analysis to ongoing optimization.

Both ASTD and ISO/IEC approaches at the same time share something and differ with the eLab approach. What they share is the focus on the teaching and learning process (Instructional Design for ASTD, process for ISO/IEC and field test for the eLab) as different from technological products. The main difference lies in the scope of the evaluation: both ASTD and ISO/IEC standards are conceived for being applied to a complete course in order to *certify* quality; for this reason they can require more resources. The eLab approach on the other hand is designed to accompany and *support* the design and development process, in order to provide guidance and, when needed, redirection.

This short comparison allows identifying some critical concerns of the eLab method. It is surely limited, mainly in extension (it does not consider a systemic view of e-Learning as the ISO/IEC method) and in precision (it does not provide structured data collection tools or procedures). Its main merit is on a different level: that of making quality evaluation doable and at the same time useful, providing enough feedback in order to make sense to project teams. Finally, certification and support are not mutually exclusive. Rather, they can be fruitfully integrated in a wide quality management program.

Conclusions

The opening of the chapter described the raising concern for quality as a sign of the growth and spread of e-Learning. We have also briefly discussed the fuzziness of the concept of quality – already discussed in other chapters in this handbook – and proposed a definition of three levels. This definition served as basis for the development of a lightweight method that supports the integration of e-Learning quality evaluation in the day-to-day practice of e-Learning design, development and delivery. The method was presented in detail, illustrated by a case study and confronted with other approaches.

Quality evaluation is a work in progress: in general needs evolve, and tools should move with them; in each single project, peculiar features should be taken into account, and the method, or the standard, adapted and reinvented each time. Quality is a challenge, and, as we have seen, a key issue for the real integration of e-Learning in our culture of education. Making it approachable and useful for practitioners is a step in this direction.

References

ASTD (2004). *E-Learning Courseware Certification (ECC).* ASTD Certification Standards Committee.

Barnett, R. (1992). *Improving Higher Education. Total Quality Care.* Buckingham, UK: SRHE & Open University Press.

Bates, T. (2001). *National strategies for e-learning in post-secondary education and training.* Paris: International Institute for Educational Planning, UNESCO.

Botturi, L. (2006). E2ML. A visual language for the design of instruction. *Educational Technologies Research & Development, 54*(3), 265-293.

Botturi, L. (2005). Object-oriented Instructional Design: What Are Learning Objects for a Higher Education Blended-learning Project? *AECT 2005 Convention*, Orlando, CA, USA [poster presentation].

Botturi, L., Cantoni, L., Lepori, B., & Tardini, S. (2006). Fast Prototyping as a Communication Catalyst for E-Learning Design. In M. Bullen & D. Janes (eds.), *Making the Transition to E-Learning: Strategies and Issues*, Hershey, PA: Idea Group, 266-283.

Cantoni, L., & Esposito, A. (eds.) (2004). *La qualità nella gestione dei progetti di eLearning nelle università italiane.* Milano, Italy: Università degli Studi.

Cantoni, L., Lepori, B., & Succi, C. (2003). European Universities managing quality in eLearning. In A. Szücs, E. Wagner & C. Tsolakidis (eds.), *The Quality Dialogue. Integrating Quality Cultures in Flexible, Distance and eLearning*, 291-298.

Cantoni, L., & Succi, C. (2003). Gestione dei processi di eLearning nelle università svizzere ed europee. *Proceedings of Didamatica 2003. Informatica per la Didattica.* Genova, Italy, 9-16.

Coimbra Group (2002). *European Union Policies and Strategic Change for eLearning in Universities.* Brussels, EU.

Collis, B., & Van der Wende, M. (2002). *Models of technology and Change in Higher Education.* Enschede: CHEPS, Toegepaste Onderwijskunde.

De Boer, H., Huisman, J., Klemperer, A., van der Meulen, B., Neave, G., Theisens, H., & van der Wende, M. (2002). *Academia in the 21st Century. An analysis of trends and perspectives in higher education and research.* Den Haag, The Netherlands: Advisory Council for Science and Technology Policy (AWT).

Ehlers, U. (2004). Quality in E-Learning. The Learner's Perspective. *European Journal of Vocational Training*, 2004/I. Retrieved on November 8th, 2006 from http://www.eurodl.org/materials/contrib/2004/Online_Master_COPs.html.

eLab (n.d.) *eLab Website.* Retrieved on November 8th, 2006 from http://www.elearninglab.org.

Hear and See (n.d.). *Hear and See project website.* Retrieved on November 8th, 2006 from http://www.hearandsee.ch.

ISO/IEC (2005) *Information technology - Learning, education and training - Quality management, assurance and metrics*, Geneva, Switzerland: ISO.

Kirkpatrick, D. L. (1994). *Evaluating Training Programs: The Four Levels.* San Francisco, CA: Berrett-Koehler.

Lepori, B., Cantoni, L., & Succi, C. (2003). The introduction of eLearning in European universities: models and strategies. In M. Kerres & B. Voss (eds.), *Digitaler Campus. Vom Medienprojekt zum Nachhaltigen Medieneinsatz in der Hochschule*, Münster, Germany: Waxmann.

Lepori, B., & Succi, C. (2004). eLearning in the Swiss Universities of Applied Sciences. *ePaper NewMine, 4*, Lugano, Switzerland: NewMinE Lab. Retrieved on November 8th, 2006 from http://www.newmine.org/NewMinE_ePapers4.pdf.

Maassen, P. A. M. (1997). Quality in European Higher Education: recent trends and their historical roots. *European Journal of Education, 32*(2), 111-127.

Massy, J., (2002). *Quality and eLearning in Europe. Survey Report 2002.* Twyford, Reading, UK: Bizmedia.

Morrison, G. R., Ross, S. M., & Kemp, J. E. (2003). *Designing Effective Instruction* (4th edition). NJ: Wiley & Sons.

Phipps, R., & Merisotis, J., (1999) *What's the Difference? A Review of Contemporary Research on the Effectiveness of Distance Learning in Higher Education.* Retrieved on November 8th, 2006 from www.ihep.com.

Phillips, J. J. (1996). How much is the training worth?. *Training and Development, 50*(4), 20-24.

Phipps, R., & Merisotis, J., (2000). *Quality On The Line. Benchmarks For Success In Internet-Based Distance Education.* Retrieved on November 8th, 2006 from www.ihep.com.

SVC (n.d.). *Swiss Virtual Campus website.* Retrieved on November 8th, 2006 from http://www.virtualcampus.ch.

Triacca, L., Bolchini, D., Botturi, L., & Inversini, A. (2004). MiLE: Systematic Usability Evaluation for E-learning Web Applications. *AACE Journal, 12*(4) [Online journal, Outstanding Paper Awards - Invited Paper from EDMEDIA 2004].

Triacca, L., Inversini, A., & Bolchini, D. (2005). Evaluating Web Usability with MiLE+. *Proceedings of the Web Site Evolution IEEE Symposium*, Budapest, Hungary, 22-29.

Van Damme, D. (2002). Quality Assurance in an International Environment: National and International Interests and Tensions. Paper presented at the *CHEA International Seminar III*, San Francisco [Background Paper].

Van der Wende, M., & van de Ven, M. (2003). *The use of ICT in Higher Education. A mirror of Europe.* Utrecht, the Netherlands: LEMMA.

Vroeijenstijn, A.I. (1995). *Improvement and Accountability: Navigating between Scylla and Charyddis. Guide for external Assessment in Higher Education.* London, UK: Jessica Kingsley Publisher.

Appendix: Acronyms and Web References

EQO: European Quality Observatory, www.eqo.info

SEEL: Supporting Excellence in E-Learning, www.seelnet.org

Qual E-learning, http://www.qual-elearning.net/

SEEQUEL: Sustainable Environment for the Evaluation of Quality in E-Learning, http://www.education-observatories.net/seequel/index

ADEC: American Distance Education Consortium, www.adec.edu

AFT: American Federation of Teachers, www.aft.org/

ASCLITE: Australasian Society for Computers in Learning in Tertiary Education, www.tlc.murdoch.edu.au/project/cutsd01.html

ASTD-ECC: The ASTD Institute E-learning Courseware Certification (ECC) Standards, http://workflow.ecc-astdinstitute.org

IHEP: Institute for Higher Education Policy, www.ihep.org

EFMD European Foundation for Management Development, http://www.efmd.org

EFQUEL: European Foundation for Quality in eLearning, http://www.qualityfoundation.org

Authors

Luca Botturi holds a Ph.D. in Communication Sciences and Instructional Design from the University of Lugano, Switzerland. He is currently instructional designer at the eLearning Lab, and researcher for the NewMinE Lab - New Media in Education Lab at the Università della Svizzera italiana in Lugano (Switzerland). He is also founder and project manager of seed, a non profit organization promoting the development of a culture of educational technologies in international development. His research interests include effective course design with new technologies and team communication.

Lorenzo Cantoni graduated in Philosophy and holds a PhD in Education and Linguistics. L. Cantoni is professor at the University of Lugano (Switzerland), School of Communication Sciences, where he teaches eLearning. He is also vice-director of the Institute of Institutional and Educational Communication, director of the laboratories webatelier.net: production and promotion over the Internet, NewMinE Lab: New Media in Education Lab, executive director of TEC-Lab: Technology Enhanced Communication Lab and co-director of eLab: eLearning Lab.

Chiara Succi graduated in Communication Sciences at the Università della Svizzera italiana works as a scientific researcher at the NewMinE Lab, where she collaborates to projects, coordinates the promotion of events and cares public relations with companies and partners. From May 2003, she is enrolled in a PhD course centred on eLearning Acceptance within companies.

Alessandro Inversini has a Master in Communication Sciences and Communication Technologies from the University of Lugano, Switzerland. He currently works as researcher at TEC-LAB in the same institution, where he is pursuing a PhD. His research activity primarily deals with usability, quality and new technologies of communication in cultural heritage tourism, with a focus on qualitative design solutions for technology-supported communication in cultural tourism. He also works as instructional designer at the eLearning Lab where he works on three Swiss Virtual Campus projects.

Buzzetto-More, N. A. (Ed.). (2007). *Advanced Principles of Effective e-Learning*
Santa Rosa, California: Informing Science Press. (pp. 185-195)

Chapter 7

Learning Lost Along the Way? Warning Signs for the E-Learning Institution

Susan Smith Nash

Introduction

This chapter details the risks and likely problems encountered when e-learning organizations grow, evolve, or change. In addition to providing tools for evaluating the health of the organization, detecting warning signs, and describing likely underlying causes, the paper suggests action steps that can be taken to bring about a rapid, sustainable, and cost-effective return to the primary goal of providing education and assuring that students can demonstrate the fact that they have achieved their learning objectives. The chapter advocates a renewed emphasis on learning outcomes and institutional vision, and describes effective leadership strategies which can allow an institution to make the changes needed in order to achieve its stated goals and provide a high-quality educational experience for its students who are taking courses at a distance via e-learning and mobile learning.

By the time an education provider realizes that it has lost its focus on learning, it may have already suffered severe, possibly irreparable damage. This a danger faced by almost every educational institution, as it strives to accommodate new demands for flexible delivery technologies, trendy curriculum and course content, and student-friendly financing options. Consequences can range mild from chaos and lack of direction to total collapse. Damage can result from faculty and staff disenchantment and low morale, a massive out-migration of students as the institution fails to live up to the promised educational vision, and financial pressure as tuition revenues fail to flow in, and donors refuse to continue to contribute to the organization.

The loss of vision is almost a natural outgrowth of rapid change, particularly if the institution's leadership has been in flux. As education providers scramble to meet the needs of their constituencies, an unconscious shift of emphasis can occur. Rather than making sure that the stated learning outcomes are the outcomes that are really being achieved, production and delivery milestones take center stage. Statistics gathered on enrollment, revenues, course completion, support services utilized, and even learner satisfaction can be hard to interpret, and provide misleading results.

Sometimes the statistics do not really provide an accurate sense of whether or not students are emerging with educational outcomes that connect with the institution's mission, vision, or objectives. The education provider easily loses sight of its own sense of identity as production-centered activities keep day-to-day operations mired in details.

Warning signs that an education provider has lost focus on learning are many and complex. For simplicity, this chapter narrows them down to a "top ten" list.

Warning Signs

The symptoms of a loss of vision and focus manifest themselves in many ways within the organization and among the students, faculty, administrators, external review boards, service providers, and other stakeholders.

Mission Measured in Numbers

One of the surest signs of a learning provider that has lost its way is a focus on enrollment and revenue numbers, without any regard to the quality of the outputs that accompany the numbers. This is a negative sign in any organization -- from producing widgets, to providing service for widgets. However, when the "production" involves people's lives and the futures of their family, community, and coworkers, to disregard the qualitative aspects of desired outcomes results in ethical conflicts. People matter, and thus a student-centered approach with a focus on the "whole person" should be maintained.

Perhaps one of the most influential theorists and educational leaders, John Dewey (1897), expressed it best in his work, "My Pedagogical Creed," which subsequently influenced generations of educators and visionaries. Education must concern itself with learning outcomes -- the tangible proof of concept, measurable in human terms. A word of caution must be applied here, however. It is a dangerous step to connect learning outcomes to employment, since in times of economic downturn, not every graduate may immediately find employment.

Focusing on numbers and on increasing enrollment may not seem to be a problem in the e-learning organization. Conventional wisdom holds that if the courses are standard and are housed in a learning management system such as Blackboard, and if advising is automated via a web interface, then it should not be difficult to accommodate more students. Administrators often suggest that increasing enrollment is just a matter of cloning more courses and buying more server space. They claim that there is no ceiling to growth, and that their model is infinitely scalable.

An institution that is in financial trouble may look to e-learning as a good way to turn around negative cash flow. They may institute "short cuts" that include credit for prior learning, or shoddily conceived mobile learning. However, expedient "short cuts" toward a degree are seductive yet dangerous. They can lead an institution down the slippery slope toward being perceived as a diploma mill (Butler, 2005). Perhaps the best antidote for the temptation to compromise standards is to keep in mind that the reputation of the organization often rests on the quality of the curriculum, instruction, and results. Constant vigilance and alignment with accrediting agency standards are vital.

Solutions to the problem of losing sight of academics in the quest for numbers can include developing a set of best practices for e-learning and requiring departments to adhere to them.

Other solutions involve establishing productive partnerships with content providers and other institutions. For example, CyberLearning is one of many companies that offer off-the-shelf courses that are offered online. What makes CyberLearning unique, however, is the fact that a full academic year of on-demand live mentoring is available and is included in the cost of the courses.

Growth for Growth's Sake

Indiscriminate growth and the pursuit of revenue without any connection to the institution's vision or mission can lead to a complete loss of focus and direction. By focusing on numbers, it is possible to substitute expediency for solid planning. For example, an institution may sign an articulation agreement with another institution without analyzing whether or not the students who transition from one institution to another are really capable of succeeding. If the students who are accepted to a college because of the articulation agreement lack sufficient preparation, it is important to provide tutoring, mentoring, and other student success services.

The e-learning organization has an advantage because of its flexibility which results in the ability to provide specific needs-based content on-demand. The problem of students who lack sufficient preparation can be solved by developing remedial activities, units, and stand-alone modules that can be delivered via e-learning or via mobile devices such as hand-held computers. Interactive grammar exercises, for example, can be delivered on handheld computers and smartphones. Math concept review can also be delivered on-demand and 24-7 via the Internet or mobile devices.

Vision Means Different Things to Different People

A sign that the institution's vision is not being implemented on a grassroots, or department level, is confusion about what the vision means. One manifestation is a failure to be consistent about what constitutes the vision. If one asks ten departments to define their institution's vision, one should expect some uniformity of response.

With e-learning, vision requires an understanding and articulation of how the technology will be used to attain learning objectives and outcomes. Further, in e-learning, vision means expressing in a concrete way just how the program will make sure that there is alignment between learner preferences and learning styles and delivery mode.

A solution to vision confusion or "blurred vision" at the e-learning institution is to develop best practices, and to make a demo or working prototype so that stakeholders can see the vision in action.

Implementing solutions can involve scheduling retreats or virtual meetings in which all administrators, staff and faculty log in and discuss the vision and the way it is being brought to life.

Cookie-Cutter Courses and Delivery

Assembly-line production without building in flexibility to accommodate diverse learner needs or access issues is a sure sign of an emphasis on quantity rather than quality. A "one size fits all" approach to delivery leads to a rigid approach to content as well as learners.

In addition to respecting cultural diversity and the way it translates into differing work styles, values, and interpersonal relations (Hofstede, 1997) where high-context and low-context cultures make it necessary to adjust instructional materials as well as instructional philosophies, it is important to accommodate differing learning styles. An understanding of multiple intelligences (Gardner, 1983) and how they translate to an online environment can make the difference between effective and ineffective courses.

While e-learning organizations may need to standardize their courses and their delivery in order to conform to the requirements of the learning management system, it is important to build in flexibility and some customizability. For example, a standard course shell that has been built at the department and approved by the office of instructional support services and technology should allow professors to add discussion threads, articles, and comments / lecture notes.

My Way or the Highway

Autocratic decision-making and the emergence of an environment that cannot tolerate an open discussion of dissenting opinions or approaches can indicate that the institution is in panic mode. In this case, the organization may be struggling to meet production deadlines, rather than seeking innovative and learner-responsive ways to assure the attainment of learning objectives.

A reasonable response to autocratic management styles can be found in a combination of transformational and transactional leadership styles (Bass, 1997). By combining the practical and measurable with the visionary, members of the organization are able to engage in productive discussions.

In the e-learning organization, the problem can be avoided. One strategy can be to make sure to offer cross-training and to make decisions using cross-functional teams. Following best practices while maintaining flexibility and giving individuals the ability to modify content to meet needs is an excellent strategy.

Reactive Rather Than Proactive Decision-Making

An education provider that feels pressured, out of control, or without clear direction can start to take on a "cornered rat" persona. A cornered rat does not listen. A cornered rat digs frantically and tries to make an escape tunnel. Failing to escape, the rat turns and bites. In an institution, instead of thoughtfully considering suggestions or making decisions in accordance with a strategic plan, people will react to the immediate situation. They will make survival-type decisions that are to take effect immediately, even when they counter the overall mission and vision of the organization.

In the e-learning organization, decisions involving infrastructure, hardware, and software are often made under pressure, either of budget constraints or because of the urgency of the need. Unfortunately, the effect can be to lock the organization into a solution that does not really meet future needs. For example, an organization may decide to implement an open-source learning management system, rather than a commercial solution. The decision seems prudent and cost-effective at first. However, there are hidden costs to contend with, and once the decision has been made to go with the open-source solution, it is necessary to pay the price for servers, support, maintenance, training and other unexpected costs. A way to avoid the problem is to interview other organizations that have made similar decisions. Even though each organization is different, it is useful to learn from the experience of others.

Cult of Personality Subsumes Democracy

A clear warning sign that an education provider has moved away from thinking of itself as a learning organization centered around the needs of a constantly evolving base of learners, is the presence of petty tyrants and autocrats who dominate through coercion or intimidation. They gained their power through fear. What kind of shock resulted in widely held organizational fear? What causes deep-seated anxiety and paranoia to be instilled in the hearts of individuals? Sometimes, the cause is technological change, resulting in an organization that can no longer achieve its original mission. Other factors include financial uncertainty, changing markets, and competition.

Fear, panic, and the desire for a strong personality to become a rescuer or savior may emerge when the organization is threatened with extinction. Instead of becoming more innovative, many individuals will react by retreating. Many seek stability at any cost. They will cheerfully give up their personal autonomy in exchange for a promise of security. What results is a closed system, where a charismatic leader exploits the insecurities of individuals within the organization (Post, 2004).

Needless to say, the organizational focus goes inward rather than outward, and the concept of a forward-looking vision collapses. The tendency is to isolate oneself and try to protect the resources one has rather than recommitting oneself to the original mission: using technology to fulfill learners' and society's educational needs. If the organization sounds eerily like a doomsday cult, take heed. The dynamics are not dissimilar.

Because the e-learning organization requires tight teamwork between technology, the academic departments, and administration, it is less likely that it will be dominated by a cult of personality. However, an autocratic leadership style can emerge, especially for the for-profit e-learning organization founded by a charismatic entrepreneur. Such situations are hard to combat. Thankfully, if the individual members of the organization share the same vision and sense of purpose, it is possible to delegate and share tasks. Establishing independent board of directors or trustees can also go a long way in avoiding autocratic leadership styles.

Lots of Queens, But Few Worker Bees

When an educational provider starts to make huge structural changes, such as adding departments, degree programs, and curriculum, or to radically overhaul delivery methods, the institution must add support staff and technical specialists at a higher rate than it adds administrators. It is not romantic, but it is a reality. Ironically, institutions faced with the need for rapid change often attempt to do it on the cheap, or to beef up the ranks of administrators.

Conventional wisdom has held that one can always count on armies of graduate assistants, interns, and temporary contract labor for the lion's share of the work. However, the e-learning organization requires workers with highly evolved technical abilities, and the learning curve is steep. To create more vice presidents while the course offering catalogue increases in girth, or a few hundred courses, is not a strategy that works. Before bowing one's head in shame, though, it must be noted that everyone has made this mistake at least once.

The e-learning institution will find that it lacks the depth and breadth of low-cost temporary labor (student workers) of a traditional college, even though it had that model in mind when it created its new degree programs. It is not unusual for an institution in such straits to seek to outsource, which is never low-cost, even though initial bids may seem so. In either case – whether the institution uses its own graduate students or interns, or turns to outsourced labor -- what can result is an academic sweat shop.

One way to avoid the sweat shop or the inappropriate use of skilled professionals is to begin to develop cross-functional teams (Maccoby, 1997) and to envision future directions and receive training on appropriate technologies and software. Teams should be comprised of individuals in different units within the organization.

Outsourcing is often a viable option, as well. Academic tasks such as tutoring can be outsourced. Strategic partnerships with content providers and other learning organizations can also lead to solid curriculum, high quality instruction, and manageable growth.

"Square Peg in Round Hole" Retrofit Planning

A clear red flag that things are amiss in the educational institution appears when learners are presented with programs inappropriate to stated organizational objectives. On an individual course level, learning objectives may be incongruent with programmatic goals. This kind of misalignment does not happen overnight, but happens when programs, delivery methods, and content are developed in blissful ignorance of the true needs of learners.

Instead of finding out how the organization's mission and vision can accommodate change, and developing a plan to creatively evolve the educational programs to attain desired learning outcomes, units find themselves repackaging old programs and content. No one -- least of the learner -- is fooled.

Alternatively, the organization may purchase off-the-shelf courses in order to make new programs available immediately. Without fine-tuning, the courses rarely express the unique character or vision of the organization. They are simply too generic. Thus, it is important to take the time to customize off-the-shelf courses and to give them the look and feel of the organization.

One way to avoid the problem is to conduct needs assessments of current and prospective students, and to conduct market analyses as a part of the overall planning process.

Another effective approach is to use database marketing to profile niche markets and to truly understand the potential new learners. Then, the goals of the organization should be re-evaluated to assure that they are, in fact, aligned with each of the courses.

Emergence or Hardening of Institutional Silos

Flexible learning requires shared resources. Although units may want to safeguard their budgets by hoarding information and/or resources, and by creating silos, ultimately this strategy is counterproductive. When storing grain, a silo is ideal because it keeps the valuable resources walled away from pests, thieves, and the elements. It allows allows direct and exclusive use of a resource in a specific production line or process. However, in organizations, a silo refers to resources, programs, and expertise that are kept separate from other units so that they will be used only for the creation of their own unit's products. A silo mentality exists when units keep all the inputs they need for their own production line, and do not use shared resources, nor do they offer to share their own.

Silos and a silo mentality will doom an organization. Instead of creating silos, education providers should be thinking of appropriate and innovative ways to partner, share, or forge associations that will result in symbiotic, mutually beneficial and sustaining relationships, which lead clearly to a way to meet learner and programmatic needs.

Granted, silo problems can be severe. The institution can seem hopelessly doomed. However, there is definitely hope. The solution is to bring in leadership that acknowledges collaborative work, with "coordinated autonomy" between units, unified by a renewed vision. Leadership must emphasize accountability and individual adherence to the institution's mission and vision, and it must reward creative ways of bringing the unit's activities in line with the institutional vision and mission. It is very helpful to map activities to outcomes, and to diagram when and where activities are failing to lead to desired learner goals and institutional outcomes.

Transformational leadership strategies (Burns, 1978) can be helpful in overcoming siloes. If the individuals who are currently in a silo are able to be united by means of a shared, unifying vision, they are more likely to start to share resources and to work together for creative solutions.

Avoiding silos can occur if individual units are motivated to share resources because it is in their best interest and there is a win-win situation. Thus, it is important to develop and implement a plan for maintaining accurate inventories of resources. If redundancy is found, the solution should involve finding ways to share resources and to partner across the institution.

How Do We Heal Ourselves? Appropriate First Steps

There are a number of ways to proceed once the fact that there is a problem has been confirmed. The first steps involve pinpointing the problems, realigning the organization with the vision and mission statements. At times, it is necessary to move back a step and make certain that the vision and mission of the organization are still appropriate. If not, it is necessary to adjust them before proceeding further. Once the vision and mission statements have been clarified and placed within a contemporary context and realistic situation, it is appropriate to assure buy-in across the organization by making sure that each unit has a voice and authority in the decision-making processes.

Change can be difficult, and the realignment of the organization can be painful, particularly when it means reallocating resources or eliminating certain programs or functions. Nevertheless, with an open mind and a willingness to train individuals to assume new responsibilities that are needed for future growth, a strong team-based program of growth can emerge, resulting in a sustainable, flexible organization.

References

Bass, B. M. (1997). Does the transactional-transformational leadership paradigm transcend organizational and national boundaries? *American Psychologist, 52*, 130-139.

Burns, J. M. (1978) *Leadership*. NY: Harper & Row.

Butler, D. W. (2005). Ivory tower rip-offs: How online diploma mills work. *e-Learners.com* www.elearners.com/resources/diploma-mills.asp. Accessed March 9, 2006.

Dewey, J. (1897). My pedagogical creed. *The School Journal, 14* (3), 77-80.

Gardner, H. (1983). *Frames of mind.* New York, NY: Basic Books.

Hofstede, G. (1997). *Cultures and organizations: Software of the mind.* London: McGraw-Hill.

Maccoby, M. (1997). Building cross-functional capability: What it really takes. *Research Technology Management. 43* (5), 56-58.

Post, J. (2004). Leaders and their followers in a dangerous world. Ithaca, NY: Cornell UP.

Author

Susan Smith Nash is Associate Dean, Liberal Arts, Excelsior College, Albany, New York. She develops and administers programs for distance delivery in multiple formats. Her research interests include leadership, new instructional technologies, teaching and learning for K-12, undergraduate, graduate, and certificate and training programs using appropriate distance technologies. Email: susan@beyondutopia.com

Buzzetto-More, N. A. (Ed.). (2007). *Advanced Principles of Effective e-Learning*
Santa Rosa, California: Informing Science Press. (pp. 197-212)

Chapter 8

Evaluating e-Learning for Quality Assurance

Dr. Nicole A. Buzzetto-More and Kaye Pinhey

Introduction

Course websites are learning objects that when used in conjunction with sound pedagogy, learning outcomes, and content can support traditional, authentic, and alternative learning and assessment protocols (Bennett, 2002). The adoption of standards and guidelines for the design and evaluation of learning objects is an important means of quality assurance (Valarmis and Apostolakis, 2006; Krauss and Ally, 2005; and Friesen, 2005) that supports the communication of meaningful feedback to instructional designers for product improvement (Krauss and Ally, 2005). In order to ensure that learning objects that support fully online instruction are well developed, a set of standards has been developed by the Office of Instructional Technology at the University of Maryland Eastern Shore that are supported by an evaluation rubric. The goal of this chapter is to encourage the establishment of guidelines and a method for evaluating e-learning quality by providing a model that can be adapted and adopted by interested institutions.

A learning object has been defined by Cohen and Nycz (2006) as a knowledge based object that is self-contained and reusable and Gallenson, Heins, and Heins (2002) refer to learning objects as a unit of instructional content that facilitates content mastery and links to learning outcomes.

Sales and Ellis (2006) explain that learning objects frequently include a variety of multimedia learning materials. According to Nash (2005) the effectiveness of learning objects is based on function, flexibility, and ability to satisfy learning objectives and adds that before a learning object can be developed learning objective, measurable learner outcomes, student levels, and content range must be established.

Online courses and online course components can serve as powerful learning objects (Nash, 2005) that when well developed and instructionally sound have proved to be effective at delivering instruction and linking to student outcomes assessment in a format that can be reused and refined over time (Buzzetto-More, 2006).

According to Cohen and Nycz (2006) e-learning has economic and social benefits to society. Additional benefits of learning through learning objects are:

- The implementation of active learning strategies (Parish, 2004)
- Solutions for individualizing instruction (Nash, 2005)
- Replicability and accessibility (Nash, 2005)
- Ease of learning management.

Since learning objects are self contained units of instructional content in e-learning they may be found as units of study located on a course website that a used either in a web-assisted or a hybrid course. A web-assisted course that is a fully classroom orientated course with a course website used to deliver learning materials, facilitate communications, and/or provide links to resources. The hybrid model blends face-to-face interaction with online learning (Buzzetto-More and Sweat-Guy, 2006). A course website in its entirety can become an effective learning object when it is part of e-learning which has been defined by Koohang and Harmon (2005) as learning experiences delivered electronically that include all learning materials and activities relating to instruction.

A number of instructional technologist support the standardization of e-learning as a means of quality assurance (Valarmis and Apostolakis, 2006; Krauss and Ally, 2005; and Friesen, 2005). Valarmis and Apostolakis (2006) explain that reusability and interoperability of content and technology is crucial. They advocate the adoption of a set of global standards for e-learning.

Krauss and Ally (2005) conducted a study that looked at how learning theories and understanding of human cognition impact the development of learning objects and then experimented with an instrument for assessing the quality of a learning object. They used eight criteria that were distributed as Likert scale survey questions to both teachers and students following their use of a specific learning object. The eight points included: content quality, learning goal alignment, feedback and adaptation, ability to motivate learning, presentation design, usability, reusability, and supporting resources. As a result of their work, they made two recommendations: 1) that more effort should be spent on the identifying and sharing of best practices for designing and applying learning objects to instruction and less time spent on the content itself; and 2) that evaluation of learning objects become a more commonplace practice.

Friesen (2005) examined the e-learning standards established by the IMS Global Consortium, the IEEE LTSC (Institute of Electrical and Electronics Engineers, Inc. Learning Technology Standards Committee), and the ISO/IEC (International Standards Organization/International Electrotechnical Commission) and noted that they primarily relate to issues of metadata and infrastructure rather than pedagogy. He explained that the difficulty in establishing standards for e-learning is that the technology is consistently evolving.

The popular Seven Principles of Effective Teaching: A Practical Lens for Evaluating Online Courses. (2001) written by Graham et al. has guided the development of many e-learning evaluation programs. It focuses completely on teaching effectiveness rather than usability or design. The seven principles include: student-faculty interaction that is encouraged and guided, discussions and assignments that encourage student cooperation, active project-based learning, providing of prompt feedback, use of deadlines, establishment and communication of expectations, and the acknowledgement of diverse learning styles.

Founded in 1886, the University of Maryland Eastern Shore (UMES) is a historically black, 1890 land grant institution and a member of the thirteen-campus University System of the State of Maryland. The student body is approximately 3,700, with 10% of the enrollment representing graduate students. UMES is located in a rural region and is the most affordable four-year institution of higher education in the state.

In a move to stimulate the use of alternative delivery methods, the regents of the University System of Maryland instituted a policy in 2005 recommending that all students take on average 12 credits through out-of-classroom experiences and other nontraditional means. Included in the regents' definition of out-of-classroom experiences are e-learning, internships, student teaching, and a host of other activities. Diana G. Oblinger, vice president of EduCause, was cited in Lorenzetti

(2005) as saying that the Maryland system is recognizing that some online learning is an enhancement to students' higher-education learning experiences even when those students are full-time on-campus residents.

WebCT 6 has been adopted as the course management system of choice by the University System of Maryland. UMES began integrating WebCT 4.0 into their curriculum in 2001. An Office of Instructional Technology was established as a result of the receipt of grant funding to support e-learning development and delivery.

In order to assure that fully-online course websites are high-quality learning objects, and after a disappointing review of existing online course showed tremendous variations in quality, the UMES Office of Instructional Technology (OIT) developed a set of guidelines and requirements as well as a rubric [see Table 1] used to evaluate completed courses. The standards were developed by a small committee comprised of instructional technologists and teaching faculty who conducted an exhaustive review of the literature on e-learning effectiveness as well as several models for online course evaluation including but not limited to: Evalutech, MacEwan, National Education Association Guidelines, the Digital Campus model, the University of Texas Model, WebCt's Model, and the model adopted by the University of Maryland College Park. Almost all models reviewed used a rubric as part of their evaluation process. After much discussion a rubric was selected and modified.

The University has decided that all online courses must be approved as satisfactory by the OIT and be considered the property of the university to be shared within the institution. All instructors developing fully-online courses are required to meet with one or more instructional technologists representing the OIT for assistance with learning object development. The courses are evaluated twice using the rubric in Table 1, once prior to instruction and again following the courses first offering.

UMES E-Learning Standards

Objectives and Student Learning Outcomes

The instructor must establish objectives and student learning outcomes that reflect not just the content of the course but also the e-learning mode of instruction (Buzzetto-More and Alade, 2006). They must be communicated in a manner that that is clear and measurable. These must be identified in advance of creating the course and must be communicated directly to learners. To help assist faculty in the development of student learning outcomes it is recommended that Bloom's Taxonomy of Educational Objectives is used to help guide the phrasing which provides a recognized set of hierarchical behaviors that can be measured as part of an assessment plan (Harich et al., 2005). The six levels of Bloom's Taxonomy relate to cognitive growth and in ascending order include: knowledge, comprehension, application, analysis, synthesis, and evaluation (Buzzetto-More, 2006).

Instructional Activities

All instructional activities should be designed to reflect the subject matter as well as the benefits derived through e-learning (e.g. project based learning, authentic instruction, Webquests, and etcetera) (Buzzetto-More, 2006). All activities should be purposeful and thoughtfully designed to assist students in achieving the instructional objectives. They must also be scaffolded, organized, and paced meaningfully throughout the semester (Digital Campus, 2002).

Assessment

Haken (2006) explained that assessment is an integral piece to assuring that an educational institution achieves its learning goals, as well as a crucial means of providing the essential evidence necessary for seeking and maintaining accreditation. Hersh (2004) advocated the position that assessment of student learning should be considered an integral part of the teaching and learning processes as well as part of the feedback loops that serves to enhance institutional effectiveness. Ridgway, McCusker, and Pead (2004) define e-assessment as the use of electronic technologies to drive student learning assessment. Vedlinski and Stevens (2002) illustrate that technology provides new means to assess learning that will yield rich sources of data. According to Buzzetto-More (2006) the implications and benefits of e-learning on assessment are numerous and e-learning should be used to help drive effective assessment programs.

Interaction/Communication & Feedback

Interaction and feedback in online learning environments is essential to success. It is important that the instructor plan for, develop activities that support, and facilitate interaction. The ability to engage in critical discourse through e-learning endeavors has been shown to benefit the learning process (Sweat-Guy, 2007). Studies by Garrison, Anderson, and Archer (2004) found that online discourse fosters critical thinking and reflection, and Wu and Hiltz (2004) explained that asynchronous communications improved students' perception of learning. Successful online discussions can allow students to demonstrate not just content mastery but the ability to incorporate content into higher level thinking; as a result, transcripts from electronic discussions have shown themselves to be valuable assessment artifacts (Buzzetto-More, 2006).

The use of communication tools should be thoughtful and educationally relevant. The instructor must be an *active participant* in online chats and/or discussions and performance expectations must be articulated to students (Sweat-Guy, 2007). The Office of Instructional Technology has developed a rubric which can be adopted by UMES faculty for assessing student discussion/chat performance.

The instructor must be readily accessible to students and provide timely and meaningful feedback to students. It is important that expectations are established and the type of feedback a student will received must be communicated to students in advance (Digital Campus, 2002). Student grades must be made available to students individually, privately, and in a timely fashion using the supplied online grading tools.

Instructional Materials

All materials delivered must enhance learning and support learning goals and should never distract and/or detract from the learning process nor place unnecessary stress or burden on students (Digital Campus, 2002). Materials must be purposeful, accessible, include instructions, logically organized, and available to students with slow internet connections.

Layout/Interface Design

The interface must be clear, professional, *simple*, appropriate, and planned in such a way to facilitate learning (Digital Campus, 2002). The layout must be logical and navigable and not confuse the learner.

Multimedia Usage

Multimedia elements can greatly enhance the learning experience for students of varying learning styles; however, when ill planned and/or developed they can be a tremendous distraction and/or place a burden on the students (Digital Campus, 2002). As a result, they must be meaningful and accessible to students.

Course Management

Course management is a vital to the e-learning equation. Instructors **can not** be absent from their online course and must manage all aspects of learning (Sweat-Guy, 2007). Online courses take time to develop and can be more time consuming initially then their in-person counterparts. Once created; however, the course can be re-used from semester to semester and placed in a learning object repository where it is shared across the larger academic community (Salas and Ellis, 2006). The instructor should also plan on offering online office hours as well as checking the course website on a daily or near daily basis.

Evaluation Rubric

Rubrics articulate the standards by which a product, performance, or outcome will be evaluated (Aurbach, n.d.). They help to standardize assessment, provide useful data, and articulate goals and objectives (Buzzetto-More and Alade, 2006).

A rubric has been designed and implemented for e-learning evaluation (see Table 1). Its initial incarnation was heavily influenced by the work of Dawn Truelson and Michelle Fisher at the Digital Campus (2002) of California State University in Fresno California. Over time; however, this rubric has been significantly modified to fit the goal of creating fully online learning experiences that serve as meaningful learning objects.

Currently at UMES, all fully online courses require approval by the Office of Instructional Technology. Following a series of discussions a minimum score was established. The minimum standard/score for endorsement by Office of Instructional Technology has been designated as 42 out of a possible 51 points.

Table 1: UMES Online Course Learning Object Evaluation Rubric

	3	2	1	0
1. Prerequisites	Prerequisites are clearly listed in multiple areas and well explained.	Prerequisites are clearly listed.	Prerequisite are incomplete, hard to access, or poorly organized.	No prerequisites listed at this time.
2. Technology requirements	Technology requirements are listed and well explained.	Requirements listed.	Requirements are incomplete.	No requirements are listed at this time.
3. Objectives and outcomes.	Objectives and outcomes are given and written in such a way that they are clear and measurable.	Listed but may not be clear and/or measurable.	Vague and are not consistent.	None exist at this time.
4. Activities support learning.	All activities are related to student outcomes and objectives	Activities appear to support achievement.	Activities do no appear to be related to goals and/or outcomes.	No activities listed at this time.
5. Assessment	Assessment of student progress is given throughout the course measures progress towards objectives and outcomes.	Assessment measures student progress towards objectives and outcomes.	Assessment does not measure progress.	No assessment addressed at this time.

	3	2	1	0
6. A variety of tools enhance interaction.	Sustained interaction is facilitated by the use of communication tools.	Interaction is facilitated by the use of communication tools.	Little interaction is facilitated by the use of communication tools.	No interaction is defined at this time.
7. Course materials.	Materials are easily to locate, accessible, understandable, meaningful, organized logically, and developed to enhance student achievement.	Most materials are easily located, accessible, understandable, meaningful, organized logically, and developed to enhance student achievement; however, some require improvement.	Materials are either difficult to locate, require too many technical steps, hard to understand, not meaningful, poorly organized, and or irrelevant.	No materials have been included at this time.
8. Student support.	Instructor uses communication tolls to provide students with readily available and timely support that helps keeps students on target.	Instructor uses communication tolls to interact with students and provide support.	Instructor sometimes communicates with students directly.	Instructor does not communicate with students directly, meaningfully, and/or in a timely manner.
9. Frequent and timely feedback.	Instructor provides frequent and timely feedback.	Instructor provides feedback.	Instructor sometimes provides feedback.	No feedback is provided.
10. Appropriate pacing.	All learning and activities are paced in a way that is meaningful and facilitates learning mastery. Calendar tool used.	Instructor paces most activities and the schedule is listed on the syllabus.	Some activities are paced.	No evidence of pacing exists at this time.

	3	2	1	0
11. Expectations for student discussion/chat participation.	Expectations for student discussion/chat participation are well developed, consistent, and communicated clearly to students.	Participation expectations are communicated to students.	Expectations are vague and/or unclear.	No expectations exist at this time.
12. Grading	Grading is timely, accessible by students, and secure.	Grades are accessible to students and secure.	Grades are inconsistently made available to students and/or are not secure.	Grades not available to students.
13.Course content	Course content is clear, meaningful, plentiful, accessible, and supported by the instructor.	Course content is for the most part clear, meaningful, accessible, and supported by the instructor.	Course content is not always meaningful, clear, and/ or accessible.	Course content is weak and lacking.
14. Navigation	Navigation is clear, logical, and all buttons/links work.	Navigation is mostly clear and most buttons/links work.	Problems with navigation exist that may cause student difficulties.	Navigation may cause confusion.
15. Display	Color, buttons, images, graphics, and etcetera enhance the course.	Some display elements are distracting and/or slow to open.	Elements may be unrelated, distracting, and/or unnecessary.	Elements are weak, distracting, burdensome, unprofessional, inappropriate, and or sloppy.

	3	2	1	0
16. Multimedia (if appropriate)	Multimedia used is appropriate, enhances learning, accessible, professionally created, linked to course content, available in an alternative text based format, and work.	Multimedia used is appropriate, enhances learning, professionally created, but may not be linked to course content, available in an alternative format, work properly, and/or accessible.	Multimedia used may be unnecessary, not linked to content, difficult to access, and or problematic.	Multimedia is unprofessional, distracting, and/or burdensome.
17. Time Devoted	The instructor devotes the appropriate amount of time to the development of the course, timely support of student learning, and sustaining of learning. Online office hours are provided.	The instructor devotes the appropriate amount of time to the development of the course, support of student learning, and sustaining of learning.	The instructor must devote more time to this course.	Not enough time has been spent and/is allocated for this course.
18. Reusability	This course will be easy to reuse, simple to modify/ improve, serve a large potential audience, and have longevity.	This course can be reused and modified easily.	This course will serve a limited audience, require tremendous modifications, be difficult to reuse or modify, and/or may not have longevity.	This course will be almost impossible to reuse.

	3	2	1	0
Results	This course is fully endorsed and the instructor is commended for their hard work.	This course is approved with modifications.	This course requires major reworking.	This course is sub par.

Results

Prior to the adoption of the standards and guidelines and the accompanying rubric, existing online courses were evaluated by a team comprised of one faculty member and one instructional technologist from the Office of Instructional Technology. The review found that approximately 50% of the existing online courses met the new standards for acceptability. Faculty were not informed of the results of the evaluation; however, all faculty were notified that they would now be required to submit an online course request and offer their course up for a rigorous review. A copy of the rubric as well as a document explaining the standards and guidelines was distributed to all faculty.

Fortunately, these activities coincided with the move from the WebCT 4 to the Web-CT 6 LMS which required all faculty to modify their courses to fit the significantly changed LMS as well as the new standards. To assist this process, training and one-on-one consultations were, and continues to be offered, in the new LMS and standards and a newsgroup was formed for the online teaching faculty. As an incentive, a mini-grant was established that awards six faculty per semester $2,000 for the creation of courses that met the new standards.

An informal questioning of the faculty who frequently teach online indicated that the standards and guidelines increased their knowledge of e-learning as well as provided meaningful feedback. The result has been an overall increase in the quality of the online course offering at UMES with all courses now meeting or exceeding the standards for acceptability.

Conclusion

Well developed learning objects should be pedagogically sound, well presented, thoughtfully managed and supported, usable, and reusable. Establishing standards and guidelines for the design and evaluation of learning objects is a valuable means of insuring quality (Valarmis and Apostolakis, 2006; Krauss and Ally, 2005; and Friesen, 2005). Assessment of learning objects provides meaningful feedback to instructional designers that can be used to improve the quality and usability of the end product (Krauss and Ally, 2005). This paper presented a working model for assessing e-learning quality that can be adopted by other academic institutions with the goal that the standardization and evaluation of learning objects becomes a more widespread practice.

References

Aurbach, E. (n.d.). *About alternative assessment.* Retrieved February 4, 2006 from http://www.aurbach.com

Bennett, R. E. (2002). Inexorable and Inevitable: The Continuing story of technology and assessment. *Journal of Technology, Learning, and Assessment.* 1 (1). http://escholarship.bc.edu/jtla/vol1/1/

Buzzetto-More, N. (2006, March). The e-Learning and business education paradigm: Enhancing education, assessment, and accountability. *Proceedings of the Maryland Business Education Association Conference.* Ocean City, MD

Buzzetto-More, N., & Alade, A. (2006). Best practices in e-assessment. *Journal of Information Technology in Education*, 5, 251-269. http://jite.org/documents/Vol5/v5p251-269Buzzetto152.pdf

Buzzetto-More, N., & Guy, R. (2006). Incorporating the hybrid learning model into minority education at a historically black university. *Journal of Information Technology in Education*, 5, 153-164. http://jite.org/documents/Vol5/v5p153-164Buzzetto130.pdf

Cohen, E. & Nycz, M. (2006). Learning objects e-learning: An informing science perspective. Interdisciplinary Journal of Knowledge and Learning Objects, 2, 23-24 http://ijklo.org/Volume2/v2p023-034Cohen32.pdf

Digital Campus of California State University, Fresno. (2002). *Guidelines and Resources for Developing and Teaching Online Courses* [White Paper Draft]

Friesen, N. (2005) Interoperability and learning objects: An overview of e-learning standardization. *Interdisciplinary Journal of Knowledge and Learning Objects, 1*, 23-31. http://ijklo.org/Volume1/v1p023-031Friesen.pdf

Gallenseon, A., Heins, J. & Heins, T. (2002). Macromedia MX: Creating learning objects. [Macromedia White Paper]. Macromedia Inc. Retrieved 6/02/06 from http://download.macromedia.com/pub/elearning/objects/mx_creating_lo.pdf

Garrison, D. R., Anderson, T. & Archer, W. (2003). A theory of critical inquiry in online distance education. In M. Moore and G. Anderson (Eds.), *Handbook of distance education.* 113-127. New York:Erlbaum.

Graham, C. Cagiltay, K. Byung-Ro, L., Craner, J. & Duffy, T. (2001). Seven Principles of Effective Teaching: A Practical Lens for Evaluating Online Courses. *The Technology Source*, March/April 2001. Available at http://technologysource.org/article/seven_principles_of_effective_teaching/

Haken, M. (2006, January). Closing the loop-learning from assessment. Presentation made at the University of Maryland Eastern Shore Assessment Workshop. Princess Anne: MD.

Harich, K., Fraser, L., & Norby, J. (2005). Taking the time to do it right. In Martell, K., & Calderon, T. (2005). *Assessment of Student Learning in Business Schools: Best Practices Each Step of the Way.* 1 (2), 119-137. Tallahassee, Florida: Association for Institutional Research.

Hersh, R. (2004, June). *Assessment and accountability: Unveiling value added assessment in higher education.* A Presentation to the AAHE National Assessment Conference. June 15, 2004. Denver: Colorado

Krauss, F., & Ally, M. (2005). A study of the design and evaluation of a learning object and implications for content development. *Interdisciplinary Journal of Knowledge and Learning Objects, 1*, 1-22 http://ijklo.org/Volume1/v1p001-022Krauss.pdf

Koohang, A. & Harmon, K. (2005). Open source: A metaphor for e-learning. Informing *Science Journal*, 8, 76-86. http://inform.nu/Articles/Vol8/v8p075-086Kooh.pdf

Lorenzetti, J. (2005). Lessons learned about student issues in online learning. *Distance Education Report*, 9(6), 1-4.

Nash, S. (2005) Learning objects, learning objects repositories, and learning theory: Preliminary best practices for online courses. *Interdisciplinary Journal of Knowledge and Learning Objects.* 1, 217-228. http://ijklo.org/Volume1/v1p217-228Nash.pdf

Parish, P. E. (2004) The trouble with learning objects. Educational Technology, research and Development, 52 (1), 49-68

Ridgway, J., McCusker, S., & Pead, D. (2004). *Literature Review of e-Assessment.* Bristol UK: Nesta Future Lab

Salas, K. and Ellis, L. (2006). The development and implementation of learning objects in a higher education setting. *Interdisciplinary Journal of Knowledge and Learning Objects*, 2, 1-22. http://ijklo.org/Volume2/v2p001-022deSalas.pdf

Sweat-Guy, R. (Spring 2007). The Role of Interaction in eLearning. In N. Buzzetto-More (Ed.) *Principles of Effective Online Teaching: A Handbook for Experienced Teachers Developing eLearning*, Santa Rosa, California: Informing Science Press.

Varlamis, I., Apostolakis, I. (2006). The present and future of standards for e-learning technologies. *Interdisciplinary Journal of Knowledge and Learning Objects, 2*, 59-76 http://ijklo.org/Volume2/v2p059-076Varlamis.pdf

Vendlinski, T., & Stevens, R. (2002). Assessing student problem-solving skills with complex computer based tasks. *Journal of Technology, Learning, and Assessment,* 1(3). http://escholarship.bc.edu/jtla/vol1/3/

Wu, D. & Hiltz, R. (2004). Predicting learning from asynchronous online discussions. *Journal of Asynchronous Learning Networks*, 8(2), 139-152.

Note

The original version of this paper was previously published in the International Journal of Knowledge and Learning Objects available at http://www.ijklo.org

Authors

Dr. Nicole A. Buzzetto-More is the coordinator of Business Education at the University of Maryland Eastern Shore; the Assessment coordinator for the Department of Business, Management, and Accounting; as well as the Co-Founder of the UMES Office of Instructional Technology. She received doctorate and masters degrees in communication and instructional technology from Columbia University and also holds degrees from the College of New Rochelle and Marist College. As a recognized assessment and e-learning expert, she is a frequent presenter at conferences across the globe. She has authored numerous publications in referred journals and has been recognized with an award from the American Distance Education Consortium.

Kaye Pinhey was a primary collaborator on the development of the standards and guidelines. Most importantly, he was the individual responsible for the implementation of the model developed. Mr. Pinhey is the Director of the Office of Instructional Technology at the University of Maryland Eastern Shore. He holds a B.S. degree from Miami University of Ohio, an M.A. from the University of Florida, as well as certifications from Novell and Microsoft. His areas of interest include quality assurance in e-learning, instructional video, multimedia development, and synchronous online instruction.

Buzzetto-More, N. A. (Ed.). (2007). *Advanced Principles of Effective e-Learning*
Santa Rosa, California: Informing Science Press. (pp. 213-244)

Chapter 9

Using e-Learning to Drive a Student Learning Outcomes Based Assessment Program

Nicole A. Buzzetto-More, Ayodele J. Alade, Dandeson Panda

Introduction

Impacting education from early childhood through graduate studies, the assessment movement is based on standards and outcomes, measuring results, and holding educational institutions accountable for student learning. Oversight bodies and accrediting agencies are beginning to require the establishment of learner-centered outcomes that reflect the well-rounded knowledge, competencies, and abilities preferred in today's students; the alignment of curriculum to reflect the desired progression and cognitive development of learners; the collection of data that demonstrates the satisfaction of learning objectives; and the use of assessment information to inform decision making (Buzzetto-More, 2006).

The use of information technologies and e-learning strategies can provide an efficient and effective means of assessing teaching and learning effectiveness by supporting traditional, authentic, and alternative assessment protocols (Bennett, 2002). According to Vendlinski and Stevens (2002) technology offers new measures for assessing learning that will yield rich sources of data and expand the ways in which educators understand both learning mastery, and teaching effectiveness. The use of information technologies and e-learning to augment the assessment process may include: pre and post testing, diagnostic analysis, student tracking, rubric use, the support and delivery of authentic assessment through project based learning, artifact collection, and data aggregation and analysis.

The purpose of this chapter is to provide insight into the key concepts of assessment as well as the ways in which technology and e-Learning serve as the most promising mechanisms for satisfying assessment goals and objectives.

This work is based on an exhaustive review of literature, the analysis of a number of critically acclaimed assessment programs, as well as the authors' own work in the development and use of e-Learning in the design, and implementation of an assessment program at a mid-sized institution of higher education located in the rural Eastern Shore of Maryland in the United States. The program being profiled includes computerized longitudinal testing, online diagnostic testing, competitive networked simulations, rubrics, student discussion transcripts, taped presentations, and electronic portfolios.

Understanding the Assessment Movement

In K-12 education, assessment first emerged in America in the 1840's, when an early pioneer of assessment, Horace Mann, used standardized written examinations to measure learning in Massachusetts (Pearson et al., 2001). After losing momentum, the scientific movement of the 1920's propelled the use of large-scale testing as a means of assessing learning (Audette, 2005). The 1960's saw further support of standardized testing when the National Assessment of Educational Progress was formed, which produced the Nation's Report Card (Linn, 2002). But perhaps no initiative has had as broad and pervasive an impact as the American No Child Left Behind Act of 2001 (NCLB), which formally ushered us into an age of accountability. The NCLB act is a sweeping piece of legislation that requires regularly administered standardized testing to document student performance. The NCLB act is based on standards and outcomes, measuring results, and holding schools accountable for student learning (Audette, 2005). In 2006 Congress is required to reauthorize the Higher Education Act and it is predicted that NCLB will lead to changes in Higher Education Assessment requirements (Ewell & Steen, 2006).

In higher education, the first attempts to measure educational outcomes emerged around 1900 with the movement to develop a mechanism for accrediting institutions of higher education (Urciuoli, 2005). In 1910 Morris Cooke published a comparative analysis of seven higher education institutions including Columbia, Harvard, Princeton, MIT,

Toronto, Haverford, and Wisconsin. The result of the report was the establishment of the student credit hour as the unit by which to calculate cost and efficiency (Urciuoli, 2005). By 1913 accreditation in higher education had spread nation wide with the formation of a number of accrediting bodies (Urciuoli, 2005). It is important to note, that the United States is unusual in that it relies on private associations rather than government agencies to provide accreditation of academic institutions and programs.

A number of reports released in the mid 1980's charged higher education to focus on student learning (Old Dominion University, 2006). During that time, the first formal assessment group was established, the American Association for Higher Education (AAHE) Assessment Forum, formed in 1987. In 1992, accrediting agencies were required to consider learning outcomes as a condition for accreditation following a 1992 Department of Education mandate (Ewell & Steen, 2006).

Assessment experts point to pioneers of the assessment movement, Alverno College and Northeast Missouri State University, which have both been committed for over three decades to outcomes-based instruction. Kruger and Heisser (1987) who evaluated the Northeast Missouri State University assessment program found that the variety of assessments and questionnaires employed as well as the use of a longitudinal database that provides multivariate analysis makes this institution an exemplar in the effective us of quality assessment to support sound decision making.

The oldest recognized undergraduate assessment program in the United States can be found at the University of Wisconsin which has reported on some form of student outcomes assessment continuously since 1900 (Urciuoli, 2005).

The assessment movement is not limited to the United States. In the United Kingdom, the Higher Education Funding Council was established following the Further and Higher Education Act of 1992, requiring the assessment of quality of education in funded institutions. In 2004, the Higher Education Act was passed with the goal of widening access to higher education as well as keeping UK institutions competitive in the global economy (Higher Education Funding Council for England, 2005). The formation of the Europe Union has created a need for the communication of educational quality. According to Urciuolo (2005) educational discourse in Europe and the UK are becoming dominated with the terms standards and accountability which were born and have been growing within the United States for many years.

Understanding Assessment

Haken (2006) explained that assessment is an integral piece to assuring that an educational institution achieves its learning goals, as well as a crucial means of providing the essential evidence necessary for seeking and maintaining accreditation. According to Dietal, Herman, and Knuth (1991) assessment provides an accurate measure of student performance to enable teachers, administrators, and other key decision makers to make effective decisions. As a result, Kellough and Kellough (1999) identified seven purposes of assessment:

1. Improve student learning;
2. Identify students' strengths and weaknesses;
3. Review, assess, and improve the effectiveness of different teaching strategies;
4. Review, assess, and improve the effectiveness of curricular programs;
5. Improve teaching effectiveness;
6. Provide useful administrative data that will expedite decision making; and
7. To communicate with stakeholders.

Most individuals in the assessment community believe that the assessment process begins with the identification of learning goals and measurable objectives (Martell & Calderon, 2005) as well as the use of specific traits that help define the objectives being measured (Walvoord & Anderson, 1998). These traits are frequently correlated with the developmental concepts articulated in Bloom's Taxonomy of Educational Objectives which provides a recognized set of hierarchical behaviors that can be measured as part of an assessment plan (Harich, Fraser, & Norby, 2005). There are six levels of Bloom's Taxonomy that relate to cognitive growth: knowledge, comprehension, application, analysis, synthesis, and evaluation. The three upper levels of Bloom's Taxonomy -- analysis, synthesis, and evaluation -- are linked to critical thinking.

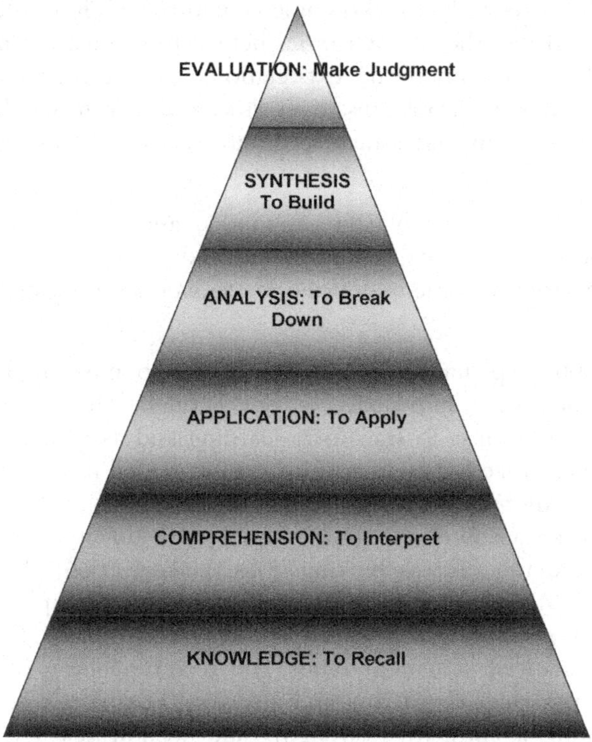

Figure 1: Bloom's Taxonomy of Educational Objectives

Figure 1 illustrates the taxonomy in its hierarchical structure.

Petkov and Petkova (2006) recommend course-embedded assessments as having the advantages of ease of implementation, low cost, timeliness, and student acceptance and note that the type of performance appraisal supported by rubrics is particularly effective when assessing problem solving, communication and team working skills. They explain that rubrics should not be considered checklists but rather criteria and rating scales for evaluation of a product or performance. According to Aurbach (n.d.), rubrics articulate the standards by which a product, performance, or outcome demonstration will be evaluated. They help to standardize assessment, provide useful data, and articulate goals and objectives to learners. Rubrics are also particularly useful in assessing complex and subjective skills (Dodge & Pickette, 2001).

Petkov and Petkova (2006) who implemented rubrics in introductory IS courses found that the use of rubrics helped to make assessment more uniform, better communicate expectations and performance to students, measure student progress over time, and help to lay the foundation for a long-terms assessment program that combines projects and portfolios.

Measuring students' knowledge, strengths, and weaknesses is done through diagnostic testing (Swearington, n.d.). Diagnostic assessment allows educators to remedy deficiencies as well as make curricular adjustments.

Haken (2006) explained that it is important to measure knowledge; however, measuring knowledge is not enough. Hence, the current charge in education is to transform learning and assessment from the world of memorized facts to a broad, well-rounded model that reflects the learner-centered outcomes of an academic program (Wright, 2004). As a result, an academic program should work on building as well as assessing students' critical-thinking skills (Haken, 2006). According to Walcott (2005), who examined business education, examples of critical thinking can be found in the creation of marketing plans, the interpretation of financial statement ratios, the recommending of organizational restructuring, identifying and analyzing ethical issues, case studies, evaluating a company's strengths and weaknesses, and portfolio creation.

Portfolios can be used to assess learning-outcome achievement as well as to diagnose curriculum deficiencies that require improvement (Popper, 2005). Popper explained that portfolios should include a variety of samples of student work. According to the American Association of Higher Education (2001), portfolios have a broad application in a variety of contexts for the collection of meaningful evidence about learning outcomes. According to Chun (2002), a portfolio should require students to collect, assemble, and reflect on samples that represent the culmination of their learning. Cooper (1999) identified six considerations of the portfolio building process: identification of skill areas, design of measurable outcomes, identification of learning strategies, identification of performance indicators, collection of evidence, and assessment.

Wiggins (1990) suggests that work being assessed should be authentic or based on the real world. Pellegrino, Chudonsky, and Glaser (2001) suggest that formative assessments focus less on student responses and more on performance. As a result, many institutions are anchoring their assessment activities into meaningful scenarios so that students are being assessed on their abilities to apply learning into realistic situations.

An assessment center is not a physical place like the name implies, but rather is a suite of exercises that are designed to replicate real life and require participants to engage in a simulation. Participants are assessed based on their real time reactions and performance (Liam et. al, 2003). Assessment Centers were first introduced in the United Kingdom in the early 1940's. They were soon adopted within the United State growing in popularity in the 1950's in corporate American where they were used by such companies as AT&T, Standard Oil, IBM, and General Electric (Liam Healy & Associates, 2003). Assessment centers have been slowly growing in higher education over the past two decades. Most commonly they are found within schools of business. They are uncommon in K-12 education; however, High Tech High School in San Diego, California has found great success with their assessment center which places learners in environments that mimic the work place and are facilitated by parents, teachers, and workplace mentors (Page, 2006).

Value-added assessment demonstrates the progress of student learning throughout a program (Martell & Calderon, 2005). It requires academics to ask "What do our students know, and how can we demonstrate that knowledge has been gained?" Value-added assessment commonly involves pre- and post-testing as well as student tracking. According to Carol Geary Schneider (2002), president of the American Association of Colleges and Universities, in order for value-added assessment to be effective it must be longitudinal, embedded within credit-bearing courses, and have weight in determining student grades.

Regardless of the assessment measures being implemented, the literature suggests that good assessment programs have variety (Swearington, n.d.). Merrimack College, for example, uses diagnostic testing, student portfolios, alumni surveys, course evaluations, rubrics, and employer surveys as part of their assessment model (Popper, 2005).

Curricular alignment occurs when a program organizes their teaching and learning activities to reflect desired student outcomes (Martell & Calderon, 2005). According to Baratz-Snowden (1993), curriculum alignment holds a school accountable for demonstrating when and where students have the opportunity to learn information and acquire skills. Engaging in curriculum alignment encourages programs to link outcomes to instruction as well as reflect upon the sequence in which competencies are built. Audette (2005) explained that curriculum alignment is particularly important to K-12 schools faced with high-stakes standardized tests. His study, conducted in the Massachusetts high school where he serves as principal, showed tangible improvement in standardized test scores as a result of curriculum alignment. Successful assessment is an ongoing cycle that involves the identification of outcomes, the gathering and analyzing of data, discussion, suggesting improvements, implementing changes, and reflection that has been depicted in Figure 2.

In Figure 2, the assessment process is represented as a continuous cyclical process or, rather, a loop. "Closing the loop" is a popular term in the assessment movement that has been defined by Martell and Calderon (2005) as ongoing process that uses assessment data to improve student outcomes. The assessment process can never be fully completed as the major goal of any successful assessment program is

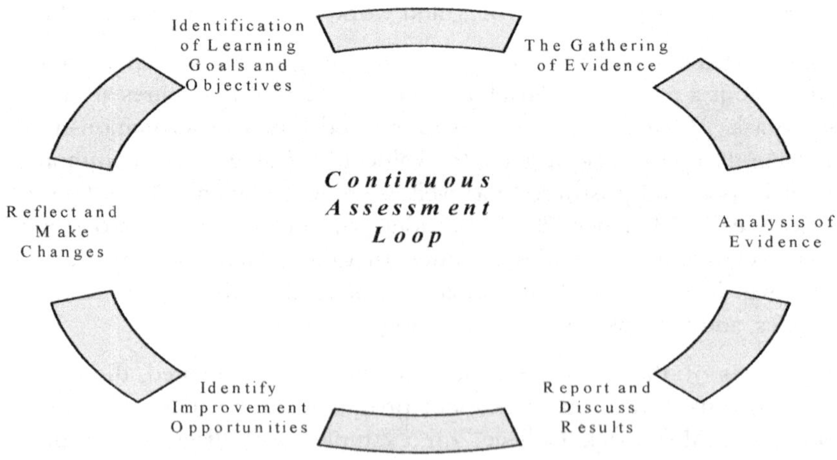

Figure 2: The Assessment Process based on Martell & Calderon (2005)

continuous educational improvement that takes into account any number of changing variables.

e-Learning and e-Assessment

Learning that is facilitated by electronic technologies; otherwise known as e-Learning, encompasses a number of activities that may or may not include the use of networked or web-based digital technologies. E-Learning strategies whether they be web-assisted, hybrid, fully online can be used to facilitate the assessment process (Buzzetto-More & Alade, 2006).

According to Bennett (2002), technology is central to learning and, as a result, is going to prove to be central to the assessment process. Bennett explains that technology will not only facilitate testing but also support authentic assessment. He refers to e-learning as part of the equipment of 21st Century scholarship and cites the success of online universities and virtual high schools in the United States.

According to the March 2006 issue of T.H.E. Journal of Technological Horizons in Education, widening achievement gaps and government mandates have required schools to use data to inform decision making, an activity that has coincided with improved information technologies, high-speed internet, and easy to use computer applications (Petrides, 2006).

Numerous studies have linked e-learning to critical thinking; for example, a study of 300 recent MBA graduates conducted at the University of Wisconsin-Whitewater found that online learning encourages high level reflective thinking (Drago, 2004).

Whereas e-learning has been part of our educational vernacular for some time, e-assessment is a fairly new term. Ridgway, McCusker, and Pead (2004) define e-assessment as the use of electronic technologies to drive student learning assessment. They say that e-assessment should encourage the rethinking of curriculum, e-learning, and technology and explain that e-assessment is flexible and supports the assessment of higher order thinking, social skills, and group work through such means as digital portfolios

Vendlinski and Stevens (2002) illustrate that technology provides new means to assess learning that will yield rich sources of data. E-assessment may include pre and post testing, diagnostic analysis, student tracking, rubric use/analysis, the support and delivery of authentic assessment through project based learning (e.g. WebQuests, simulations, eportfolios), artifact collection, and data aggregation and analysis. (Buzzetto-More, 2006).

Computerized delivery and analysis of diagnostic or traditional testing is increasing in popularity. According to Hamilton and Shoen (2005), web-based testing has significant advantages in the areas of cost, ease of use, reliability, replicability, scoring, aggregating results, and data management. They explain that digital assessment measures can score themselves with great reliability and no subjectivity while making data available with immediacy. Texas Christian, for example, requires students entering its business program to complete a computerized diagnostic test in order to collect meaningful data about incoming students (Gay, 2005).

Rubrics can be translated to a digital format where they may be made available through an intranet or over the internet. Used for scoring, these scores provide meaningful assessment information. When connected to a database, they provide educators with data that can be aggregated (Buzzetto-More, 2006). There are a number of websites that assist teachers in the development of rubrics. Two useful rubric builders can be found at http://rubistar.4teachers.org and http://landmark-project.com

As mentioned earlier, experts suggest that work being assessed should be authentic or based on the real world (Wiggins, 1990). Frequently known as project based learning, it is a form of instruction where students are immersed in challenging learning situations that are anchored in real world simulations (Page, 2006). According to Page, project based learning can support critical thinking, multilayered decision making, goal setting, problem solving, and collaboration. As a result, many institutions are anchoring their assessment activities into meaningful scenarios so that students are being assessed on their abilities to apply learning into realistic situations.

The World Wide Web contains numerous project based learning activities as well as a variety of mechanisms for their development, hosting, and or support (Page, 2006). GlobalSchool.Net is an excellent resource for gaining information on project-based learning. Another great resource is the WebQuest.

A WebQuest is an inquiry-oriented learning activity where the majority of information that learners interact with comes from web resources (Dodge, 1995). Bernard Dodge of San Diego State University, the developer of the WebQuest, explained that WebQuests give students clarity and a purpose to online queries by encouraging students to compare, classify, analyze, evaluate, and synthesize. WebQuests have been developed for all levels of education from elementary through adult education. There are variations among WebQuests; however, all should contain the following elements:

- An **introduction** that provides a narrative and background information.
- A **task** that is meaningful, realistic, and interesting.
- A set of **resources** needed to complete the task.
- A clear description of the **process** or steps for accomplishing the task.
- Some **guiding questions** or directions on how to organize the information acquired. Templates, timelines, concept maps, or cause-and-effect diagrams as described may also be included.
- An **evaluation** section that explains how performance will be assessed.
- A **conclusion** that brings closure and reminds learners about what they have gained.

Computer simulations are a form of project based learning that require learners to discover and apply learned skills interactive changing environments that mimic real-world situations (Berge 2002). Vendlinski and Stevens (2002) recommend the IMMEX multi-media exercise software, a web based simulation authoring, presentation, and assessment tool. Granland, Bergland, and Erikson (2000) reviewed three different web-based simulations at three different universities and found that the simulations encouraged critical thinking and discovery learning; however, they also noted the lack of quality assessment measures available to evaluate student simulations. As a result, educators are increasingly finding the value of using rubrics to fully evaluate simulation participation because the score or end result is not always indicative of the students thought processing and participation (Buzzetto-More & Alade, 2006).

Studies by Garrison, Anderson, and Archer (2004) found that online discourse fosters critical thinking and reflection, and Wu and Hiltz (2004) explained that asynchronous communications improved students' perception of learning. A study conducted in the United Arab Emirates indicated that students who are reluctant to participate in classroom discussions are more vocal in electronic discussions and that discussions increase understanding of course content (Bhatti, Tubaisahat, & El-Quawasmeh, 2005). Successful online discussions can allow students to demonstrate not just content mastery but the ability to incorporate content into higher-level thinking; as a result, transcripts from electronic discussions have shown themselves to be valuable assessment artifacts (Buzzetto-More, 2006).

According to the American Association of Higher Education (2001), portfolios have a broad application in a variety of contexts for the collection of meaningful evidence about learning outcomes. Portfolios are an effective form of alternative assessment that encourages students and educators to examine skills that may not be otherwise accessed using traditional means such as higher order thinking, communications, and collaborative abilities (Buzzetto-More & Alade, 2006; Wright, 2004). According to the ePortConsortium (2003) the benefits of electronic portfolios in education are that they help students develop organizational skills; recognize skills, abilities, and shortcomings; showcase talents; assess academic progress; demonstrate how skills have developed over time; make career decisions; demonstrate that one has met program or certification requirements; and promote themselves professionally.

A portfolio should require students to collect, assemble, and reflect on samples that represent the culmination of their learning (Chun, 2002) providing students with a diversity of opportunities to skills and abilities (Martell & Calderon, 2005). New High Tech High School in Napa, California involves students in long-term technology dependent portfolio projects (Page, 2006) and the business program at Merrimack College, has students engaged in an ongoing portfolio building process throughout their time in the program where students revisiting their portfolio will be able to observe their own growth (Popper, 2005).

Online portfolios have a number of advantages over those that are paper based as they support: a greater variety of artifacts and allow for greater learner expression; are dynamic and multimedia driven; accessible by a large audience; contain meta-documentation; easy to store; and may serve to promote a student academically or professionally (Buzzetto-More & Alade, 2006).

The collection of artifacts contained within a student learning portfolio can include text based items, graphic or multimedia driven elements, websites, and/or other items captured in an electronic format (Lorenzo & Ittelson, 2005a). According to Lorenzo and Ittelson (2005b) the skills required in the creation of electronic portfolios helps students learn, understand, and implement the information literacy process.

Information literacy is the ability to collect, evaluate, assemble, reflect upon, and use information in order to learn and inform problem-solving and decision making (Bruce, 2003). It is a skill crucial to lifelong learning that is dependent on the ability to engage in critical and reflective thinking (Bruce, 2003). Electronic portfolios are quickly becoming the primary means in academia for students to demonstrate and reflect on learning in a way that helps students build and apply information literacy skills (Buzzetto-More, 2006b).

Many teacher education programs have successfully incorporated e-portfolios into their curricula as a means to satisfy current National Council for Accreditation of Teacher Education (NCATE) standards. These systems frequently incorporate the use of scoring rubrics that collect statistical data for later aggregation (Barett, 2004). Popular systems include: TaskStream, LiveTech, TK20, Foliotek, FolioLive, ePortfolio, TrueOutcomes, and SpringBoard. The two biggest companies offering course management systems WebCT and Blackboard have merged and have introduced their new electronic portfolio system which is available as an add on product to their LMS.

Other technologies that are gaining in popularity in e-assessment include pen top computing (which allows teachers to review, comment, add to, and access handwritten student notes and work), integrated student response keypads (which allow for real time whole class questioning and data collection and analysis), pod casting (recording and distributing small audio and video files to students via their handheld devices), and digital video/audio lecture capturing synched with tablet pc presentations and activities (providing an archived record of teaching effectiveness for assessment demonstration). All of the aforementioned technologies not only augment the teaching and learning process but also provide data and/or artifacts that can help to satisfy assessment objectives (Buzzetto-More, 2006a).

In order to expedite the data collection and review process, many schools are choosing to automate their data collection, storage, and analysis, which has been identified as a best practice by Hamilton and Shoen (2005). Assessment data management systems are frequently web-based and made available by a university's intranet, with a system of priority based accessibility in order to ensure data security (Dhir, 2005). They may include data collected from rubrics, portfolios, student placement tests, diagnostic tests, grades, advisement information and participation in university activities, use of remediation services, technology use, attendance, and other useful information (Dhir, 2005). According to Love and Cooper (2003), assessment systems must take into account issues of interface, accessibility, security, usability, the information to be collected, hardware and software technology, and information storage and processing.

Berry College's business department employs a detailed web-based assessment management system that in addition to containing crucial assessment data also includes process maps of the various procedures that occur throughout the institution and that occur in individual schools, departments, and or offices (Dhir, 2005). The business program at Merrimack College considers their data management system crucial to their assessment process, allowing them to produce assessment reports, school and department assessment portfolios, and make information available to a variety of stakeholders with varying degrees of accessibility online (Popper, 2005).

Western Washington University is currently developing a school-wide assessment data system that will allow administrators to access pertinent assessment data through their secure intranet d and Wake Forest University which puts a laptop in the hands of all of their students also puts their course evaluations online enabling immediate analysis of evaluations as well as the ability to aggregate the data collected (National Center for Postsecondary Education [NCPE], 2001).

Northwest Missouri University uses a system of dashboards to facilitate their assessment program. Their assessment information is distributed through a networked system of linked and queriable spreadsheets that are referred to as dashboards and profiles (NCPE, 2001).

An e-Learning Facilitated Assessment Model in a Mid-Sized Business Department

After conducting an extensive review of the prevailing literature, examining a variety of assessment models, and investigating a number of projects; a business department at a four year institution of higher education in Maryland, driven by their application for accreditation by the American Association of Colleges and School of Business International (AACSB), has developed an assessment plan that they consider to be both ambitious and practical. They believe that the plan when fully implemented will result in a tangible improvement in the total teaching and learning experience. The process, which has been long in its development, was initiated with the development and articulation of a mission, goals, and measurable learner outcomes (Table 1)

The University of Maryland Eastern Shore is a public institution and the student body is approximately 3,700, with 10% of the enrollment representing graduate students. Located in a rural region, it is the most affordable four-year institution of higher education in the state. The Department of Business, Management, and Accounting is one of the largest departments on campus with approximately 430 majors, offering programs that include Business Management, Marketing, Accounting, and Business Education.

Table 1: Department of Business Management & Accounting Core Curriculum Mission, Goals, & Outcomes

MISSION & GOALS	EXPECTED STUDENT LEARNING OUTCOMES
MISSION The mission of the Department of Business, Management and Accounting at the University of Maryland Eastern Shore is to deliver high quality undergraduate management education to students majoring in business administration, accounting, and business education, as well as to provide core management courses to other majors throughout the University. This mission is supported by the intellectual contributions of faculty, in that these scholarly activities contribute to instructional effectiveness. **Goal I** To graduate students who can demonstrate a basic understanding of managerial and accounting procedures, professional practices, and issues and problems related to business and the regional and global economy.	Through a variety of experiences, graduates will be able to: • Describe major business theories and principles. • Relate group dynamics and organizational understanding to cooperative business practice. • Explain the basic tenets of accounting, marketing, and management.
Goal II To graduate students who have developed the oral, written, interpersonal, technological, and presentational skills to succeed in today's global business environment.	• Apply the proper mechanics, organization, format, style, and professional language to communicate effectively in business writing. • Employ the proper verbal, nonverbal, and presentation skills in order to communicate professionally and persuasively. • Effectively utilize technology to augment business practice: including the acquisition, analysis, and communication of ideas; and the management, organization, and examination of information.
Goal III To build upon the University's solid liberal arts grounding to foster the students' ability to integrate critical, theoretical, ethical, and global perspectives into business practice.	• Identify and describe legal and ethical issues impacting business practices. • Discuss the global nature of business. • Analyze information and synthesize concepts in order to develop and communicate ideas.

Measuring the Knowledge and Comprehension of Students

Sometimes in the assessment game, educators are able to build on a program already in existence. Such was the case with the measurement of student knowledge and comprehension. For a number of years, all students enrolled in the Department of Business, Management and Accounting have been required to take three professional development seminars at the freshman, sophomore, and junior levels. Historically, a take-home multiple-choice test has been administered to measure student knowledge and comprehension of key business concepts. The same test has always been administered in each seminar in order to track the knowledge gained during the students' progression through the program.

As part of the current assessment initiative, this test has been transferred to an online format that is delivered and graded via the WebCt course management system. Rather than remaining open-book, the exam is being delivered in a proctored computer lab, where the scores are automatically calculated and stored for comparative analysis. By tracking student knowledge and comprehension of core content as it is built during the students' progression through the program, the department is conducting value-added assessment.

Student Communications Skills

The written and oral presentation abilities of students are skill-based competencies that are crucial to the future career successes of today's graduates. Professional presentations are required of students across the program's curriculum, increasing in complexity gradually. Business writing is enhanced in a dedicated course, business communications, required of all students enrolled within the department who have previously completed the written and oral communications requirements of the university.

Business communications runs as a paperless hybrid learning experience using the WebCT learning management system. Hybrid courses blend face-to-face interaction with online learning and involve the online delivery of lectures and other curricular materials (Buzzetto-More & Guy, 2006). In order for a course to be considered hybrid, some seat time must be forfeited to accommodate the additional weight put on the online work. Recently, the use of Tegrity has been added to the course, enabling the instructor to capture key lectures with live audio synched with screen-captured video from the instructor's tablet computer. The technology used and the content of this course have enabled it to become part of two separate assessment protocols examining written and presentational communications, respectively.

Writing

Online diagnostic testing of student writing is administered at two junctures during business communications. A diagnostic test focuses on: sentence structure, subject-verb agreement, pronouns, adjectives and adverbs, verbs, as well as numerous aspects of grammar. The test is administered twice and the diagnostic results of the first test are evaluated and communicated to students within the first two weeks of the semester. Students are provided with a number of web-based resources to improve their deficiencies, and, if necessary, remediation is recommended. The second diagnostic, which occurs later in the semester, is used to assess student progress toward enhancing their strengths and rectifying their weaknesses. The test is made available through the course website and the online diagnostic system being used makes a variety of reports available for analysis.

A business-writing portfolio is also assigned and collected digitally. The written portfolio is accessed using a common rubric. Sample student portfolios and corresponding digital rubrics (created, stored, and aggregated in WebCT 6) are archived as assessment artifacts.

A writing competency exam is administered following the completion of the course and passage of the exam is a requisite to graduation. To assist in the assessment process, the department's writing rubric is being utilized. Samples of the exam are collected, without identifying information, and annually reviewed using the standards set forth in the rubric (see Table 2) by an assurance of learning committee. The committee is charged with identifying areas that require improvement and determining a course of action. Additionally, although the exam is currently paper based, the pass-fail rates of students are calculated and stored. Currently, a plan is in place to transfer the exam to a digital format where it can be administered in a lab.

Table 2: Writing Assessment Rubric

	Lacking	Developing	Satisfactory	Exemplary
Format	**Proper form not applied.**	**Attempt at formatting illustrated.**	**Acceptable formatting.**	Met all formatting specifications.
Introduction/ Thesis	**Missing.**	**Attempt demonstrated; however, it is incomplete.**	**Contains all the necessary elements; however, improvement is needed.**	Gains attention, is summative, absent of errors.
Organization/Focus	**Sequence of ideas is not logical.**	**Somewhat unorganized or unfocused.**	**Adequate, some improvement is needed.**	Well organized, focused, and flowed seamlessly.
Knowledge of Subject/ Support	**Author appeared uninformed.**	**Some attempt at understanding the topic.**	**Knowledge illustrated; improvement needed.**	The author was clearly well versed on the subject.
Mechanics/ Word Choice	**Numerous errors in spelling, grammar, usage, tone, or language.**	**Some problems with spelling, usage, tone, language, or vocabulary.**	**The work had few errors and most words were well chosen.**	The work was error free requiring no editing. Excellent work choice.
Conclusion/Follow Up	Missing	There but weak	Adequate, improvement is needed.	Clear, summative, and concluding.

Oral Presentations

Student oral presentations are being assessed using a standard rubric (see Table 3) in three courses that represent three stages of a student's progression through the department. The courses are introduction to marketing, offered at the sophomore level; business communications, offered at the junior level; and the capstone course, offered to graduating seniors. For maximum effectiveness, multiple assessors are involved.

Student presentations are being recorded on video at random. Additionally, in an attempt to further communicate assessment and collect data, the Tegrity system is being used in random sections of business communications to record student presentations using a webcam synched with instructor assessment conducted through a rubric on the instructor's tablet computer. Students are able to visit the course website and review the presentations and instructor evaluation in a split screen format. This also enables student presentations to be maintained as part of the department's data/artifact collection.

Table 3: Professional Presentation Assessment Rubric

Criteria	Weak	Developing	Satisfactory	Exceptional Error Free
Introduction	Missing	Attempt illustrated.	Adequate but needs works.	Prepares audience, summarizes.
Organization	Lacking logical sequence or flow.	Attempted but at times unorganized.	Adequate, requires improvement.	Organized, focused, and flowed.
Delivery Verbal	Poor verbal delivery.	Some areas require major improvement.	Satisfactory but improvement is needed.	Exemplary verbal delivery that is well polished.
Delivery Nonverbal	Significantly flawed and distracting.	Attempt evident. Significant improvement is needed.	Adequate improvement needed.	Professional and polished in nature.
Content	Content is weak.	Attempted needs major improvement.	Adequate but more preparation is needed.	Exemplary an outstanding effort
Visual Aid (if applicable)	Missing	Attempted but ill prepared.	Adequate; problems evident.	Supportive, created interest.
Conclusion	Missing.	Attempt evident but weak.	Adequate needs work.	Summative, signals end.

Critical Thinking

Case Studies

The ability for students to analyze situations, evaluate and make judgments, and formulate solutions are critical-thinking skills, all of which case analysis requires. As a result, the department has begun testing the use of a common rubric (Table 4) to assess case studies completed by students in two sections of business ethics. In one section of the course, students are submitting their cases in a digital drop box in WebCT, where they are archived for future evaluation. Plans are in place to adopt the rubric in other courses such as marketing where case analysis is also a commonplace.

Table 4: Case Analysis Rubric

	Weak	Developing	Satisfactory	Outstanding
Format	Incorrect format employed.	Attempt limited improvement is needed.	Well formatted with minor improvement needed.	Outstanding format.
Organization	Illogical or disjointed organization.	Attempted but improvement is needed.	Satisfactory, improvement needed.	Outstanding-logical, focused, and clear.
Content	Weak main issues not identified.	Attempt but needs improvement.	Satisfactory with only minor additions or improvement needed.	Presented all important issues & supporting materials.
Mechanics	Many errors.	Some unnecessary errors evident.	Good with minor errors.	Virtually error free.
Analysis	Weak analysis.	Cursory analysis provided that lacked depth.	Satisfactory analysis that needed only a little more depth.	All issues identified, course concepts applied, judgments made, ideas assembled.

WebQuests

Students enrolled in Business Communications are required to complete a business plan research and development group WebQuest and students enrolled in the Business Education program are required to create original WebQuests, as well as accompanying lesson plans. WebQuests are evluated using rubrics (Table 5 depicts the rubric used to assess business plan WebQuest performance. WebQuests completed by students in Business Communications as well as the WebQuests created by students majoring in Business Education are reviewed and archived as well as the corresponding rubrics.

Table 5: Business Plan WebQuest Rubric

	Weak	Developing	Good	Outstanding
Feasibility	Not feasible.	Only portions are feasible.	Some aspects not feasible.	All aspects of the plan are feasible.
Demand	No demand is demonstrated.	Demand is poorly demonstrated.	Demand is likely but could be better demonstrated.	Demand demonstrated with meaningful justification.
Communication of plan	Plan is poorly communicated.	Attempt to communicate important ideas and issues.	Satisfactory some improvement needed	Presented all important issues & supporting materials.
Mechanics	Weak	Some errors evident.	Minor errors.	Virtually error free.
Organization	Poor organization.	Attempted, major improvement needed.	Satisfactory, improvements needed.	Plan is clearly and logically organized.
Support	Not supported with findings.	Attempted but irrelevant, and/or inadequate.	Satisfactory improvements needed.	Supported by timely and relevant data.

Simulations

Simulations require critical thinking as well as the application of knowledge and skills in authentic anchored scenarios. Students in the department consistently rank nationally in competitive networked simulations, and the department considers these simulations to be an integral part of the student learning process. Simulations are done in several courses and are scaffolded through the curriculum where each simulation builds on knowledge and skills built in previous simulations. Computerized simulations can currently be found in several mandatory courses including marketing, entrepreneurship, and strategic management. The product Capstone©, available through Management Simulations, Inc., is being utlilized. It involves students fulfill the roles of the managers of competing companies, making decisions in Research and Development, Production, Marketing, Finance, Human Resources and Total Quality Management (Mitchell, 2006). A corresponding test that measures knowledge is administered at the completion of the simulation.

Beginning in the spring of 2006, simulation scores, comparative data on student performance, test scores, and a survey that examines student satisfaction and perceptions are being collected and reviewed. Additionally, because simulation results do not necessarily indicate the thought processing, knowledge application, analysis, synthesis, and evaluation being done by students, a rubric is being developed to assess students' engagement in simulations. This will help address a need that has been identified by both Lunce (2004) and Granland, Bergland, and Erikson (2000) as absent from the literature on the educational impact of online simulations. The rubric is currently under development and will be available for, and implemented in, the Spring 2007 semester.

Student Portfolios

Student portfolios are project-based learning activities that also serve as effective means of assessing learning-outcome achievement, by providing students flexible opportunities to demonstrating the acquisition of skills and abilities. An electronic portfolio has been adopted within the business education teacher education program using the TK20 portfolio assessment system. The electronic portfolios created by students include: lesson plans, WebQuests, student teaching videos, images, reflective journal entries, papers, transcripts, evaluations completed by cooperating teachers, observations made by their program advisor, student projects, rubrics, study aides, PowerPoint Presentations, web-sites, curriculum maps, goals and objectives, seating charts, behavior expectation sheets, assessment materials, progress reports, and a variety of other artifacts that demonstrate a students mastery of the principles established by the Interstate New Teacher Assessment and Support Consortium which have been adopted by the University. Portfolios are presented by the students and assessed using a simple rubric by a team of assessors. The portfolio is accessible to students for a period of seven years following their graduation from the program and has shown itself to be a useful resource for students applying for employment as it allows them to communicate a variety of skills and abilities in a dynamic format.

Impressed by the success of the capstone portfolio used in the business education teacher education program, an e-portfolio model is being introduced by the larger Department of Business, Management, and Accounting for use in all senior capstone courses. An agreement was made with the Blackboard/WebCT company to pilot their new electronic portfolio product. The portfolio has been structure around the department's student learning outcomes and it is the responsibility of the students to select artifacts that demonstrate mastery of these outcomes as well as the author meaningful explanatory reflections.

The student capstone portfolio will be assessed by multiple assessors applying standards communicated through a rubric currently under development. The e-portfolio model is preferred due to the accessibility, flexibility, and professional usefulness.

Summary and Future Work

The assessment program of the Department of Business Management and Accounting is still in its infancy and, like any infant, is bound to stumble while it makes its first steps. Noteworthy in this program is the commitment by the faculty and leadership of the department, the multiple assessment methods being employed, the use of educational technologies, and the genuine dedication to improving educational effectiveness.

The department is currently exploring a variety of assessment systems in an attempt to choose one for future adoption. The goal is to have a detailed data-management system in place that will enable faculty across the department, university administrators, and accrediting agencies to review data and artifacts on a continuous basis. The use of a multi-queriable assessment database allows the department to run an extensive variety of correlations relevant to the overall quality of teaching and learning, as well as to automate administrative functions. The data-management system under consideration will include: placement-test results, grades, advisement information, participation in university activities, diagnostic scores, rubric ratings, videos, attendance information, use of remediation services, samples of student work, and other useful artifacts. For security reasons, varying levels of accessibility will be determined based on the needs of the users, and in many instances student identifiers will be removed.

To drive the assessment process, an assurance of learning committee has been established and given the charge of insuring the success of the assessment plan and annually reviewing artifacts and data in order to pinpoint areas that require action and or improvement. The committee is the decision making source that will determine if a course of action is needed as well as responsible for developing a corresponding plan. Documentation and justification for all decisions is recorded in detail. Additionally, the changes made, the implementation of the changes, the results, impact on student learning/performance, and successes and failures are considered and used to determine future courses of action.

Lessons Learned

Extensive investigation has illustrated not just the importance of student outcomes assessment but also the critical role that technology and e-learning strategies can play in an overall assessment program. With the increasingly dependent and vital role that technology plays in human productivity and knowledge acquisition it stands to reason that technology needs to play an important role in our efforts to evaluate instruction and learning outcomes, as well as drive the decision making that seeks to enhance educational effectiveness.

Although technology facilitates assessment and e-assessment has been identified as a best practice, the most important lesson learned is that assessment programs are time consuming efforts that require planning and foresight. Effectiveness is dependent on institutional and administrative support as well as a long range plan for sustainability that includes technological preparedness.

References

American Association for Higher Education. (2001). Electronic portfolios: Emerging practices for students, faculty and institutions. Retrieved 2/28/06 from http://aahe.ital.utexas.edu/electronicportfolios/index.html

Audette, B. (2005). Beyond curriculum alignment: How one high school is using student assessment data to drive curriculum and instruction decision making. Retrieved 1/15/06 from: http://www.nesinc.com/PDFs/2005_07Audette.pdf

Aurbach, E. (n.d.). About alternative assessment. Retrieved 2/4/06 from http://www.aurbach.com/alt_assess.html

Baratz-Snowden, J.C. (1993). Opportunity to learn: Implications for professional development. *Journal of Negro Education, 62*, 311-323.

Barett, H. (2004). Differentiating electronic portfolios and online assessment management systems. *Proceedings of the SITE Conference.* Retrieved 2/15/06 from http://electronicportfolios.com/portfolios/SITE2004paper.pdf

Bennett, R. E. (2002). Inexorable and inevitable: The continuing story of technology and assessment. *Journal of Technology, Learning, and Assessment, 1* (1). Available at http://www.jtla.org

Berge, Z. L. (2002). Active, interactive and reflective eLearning. *Quarterly Review of Distance Education, 3* (2), 181-190)

Bhatti, A., Tubaisahat, A., & El-Qawasmeh, E. (2005). Using technology-mediated learning environment to overcome social and cultural limitations in higher education. *Issues in Informing Science and Information Technology, 2*, 67-76. Available at http://2005papers.iisit.org/I06f77Bhat.pdf

Bruce, C. (2003). *Seven faces of information literacy.* Retrieved 6/12/06 from: http://crm.hct.ac.ae/events/archive/2003/speakers/bruce.pdf

Buzzetto-More, N. (2006, March). The e-Learning and business education paradigm: Enhancing education, assessment, and accountability. *Proceedings of the Maryland Business Education Association Conference. Ocean City, MD.*

Buzzetto-More, N., & Guy, R. (2006). Incorporating the hybrid learning model into minority education at a historically black university. *Journal of Information Technology in Education, 5*, 153-164. Available at http://jite.org/documents/Vol5/v5p153-164Buzzetto130.pdf

Chun, M. (2002). Looking where the light is better: A review of the literature on assessing higher education quality. *Peer Review. Winter/ Spring.*

Cooper, T. (1999). *Portfolio assessment: A guide for lecturers, teachers, and course designers.* Perth: Praxis Education.

Dhir, K. (2005). Content access, and the use of data for student learning: The case of Berry College. In K. Martell & T. Calderon, *Assessment of student learning in business schools: Best practices each step of the way* (Vol. 1, No. 2, pp. 167-183). Tallahassee, Florida: Association for Institutional Research.

Dietal, R. J., Herman, J. L., & Knuth, R. A. (1991). What does research say about assessment? North Central Regional Educational Laboratory. Retrieved 3/27/06 from: http://www.ncrel.org/sdrs/areas/stw_esys/4assess.htm

Dodge, B. (1995). WebQuests: A technique for internet-based learning. *Distance Educator, 1* (2), 10-13

Dodge, B., & Pickette, N. (2001). Rubrics for Web lessons. Retrieved 2/12/06 from http://edweb.sdsu.edu/webquest/rubrics/weblessons.htm

Drago, W. (2004, July 23). Study: Virtual courses producing measurable gains in high-level learning. *Ascribe.* Retrieved 2/12/06 from http://www.ascribe.org/cgi-bin/spew4th.pl?ascribeid=20040723.102807&time=10%2050%

ePortConsortium.Org. (2003). Electronic portfolio white paper [Version 1.0]. Retrieved 4/14/06 from http://eportconsortium.org

Ewell, P. & Steen, L. A. (2006). The four A's: Accountability, accreditation, assessment, and articulation. The Mathematical Association of America. Retrieved 3/13/06 from http://www.maa.org/features/fouras.html

Garrison, D., Anderson, T., & Archer, W. (2004, May). Critical thinking, cognitive presence, and computer conferencing in distance education. Retrieved 2/12/06 from http://communitiesofinquiry.com/documents/CogPres_Final.pdf

Granland R., Bergland, E., & Eriksson, H. (2000). Designing Web-based simulations for Learning. *Future Generation Computer Systems, 17*, 171-185.

Gay, W. (2005). (Almost) painless assessment: Using intake processes for assessment purposes. In K. Martell & T. Calderon, *Assessment of student learning in business schools: Best practices each step of the way* (Vol.

1, No. 2, pp. 24-46). Tallahassee, Florida: Association for Institutional Research.

Haken, M. (2006, January). Closing the loop-learning from assessment. Presentation made at the *University of Maryland Eastern Shore Assessment Workshop*. Princess Anne: MD.

Hamilton, D. & Shoen, E. (2005). Same song, second verse: Evaluation and improvement of an established assessment program. In K. Martell & T. Calderon, *Assessment of student learning in business schools: Best practices each step of the way* (Vol. 1, No. 2, pp. 138-153). Tallahassee, Florida: Association for Institutional Research.

Harich, K., Fraser, L., & Norby, J. (2005). Taking the time to do it right. In K. Martell & T. Calderon, *Assessment of student learning in business schools: Best practices each step of the way* (Vol. 1, No. 2, pp. 119-137). Tallahassee, Florida: Association for Institutional Research.

Hersh, R. (2004, June). Assessment and accountability: Unveiling value added assessment in higher education. A Presentation to the *AAHE National Assessment Conference.* June 15, 2004. Denver: Colorado

Higher Education Funding Council for England. (2005). About us: History. Retrieved 3/13/06 from http://www.hefce.ac.uk/aboutus/history.htm

Kellough, R.D. & Kellough, N.G. (1999). *Secondary school teaching: A guide to methods and resources; planning for competence.* Upper Saddle River, New Jersey Prentice Hall.

Kruger, D. W., & Heisser, M. L. (1987). Student outcomes assessment: What institutions stand to gain. In D.F. Halpern (Ed.), *Student outcomes assessment: What institutions stand to gain. New Directions for Higher Education* (pp. 45-56). San Francisco: Jossey-Bass.

Liam Healy & Associates. (2003). Assessment and development centers. Retrieved 3/13/06 from http://www.psychometrics.co.uk/adc.htm

Linn, R. (2002) Assessment and accountability. *Educational Researcher, 29* (2), 4-16).

Lorenzo, G, & Ittelson, J. (2005a, July). An overview of e-portfolios. *EduCause Learning Initiative Paper 1: 2005*

Lorenzo, G, & Ittelson, J. (2005b, October). Demonstrating and assessing student learning with e-portfolios. *EduCause Learning Initiative Paper 3: 2005*

Love, T. & Cooper, T. (2004). Designing online information systems for portfolio-based assessment: Design criteria and heuristics. *Journal of*

Information Technology Education, 3, 65-81. Available at http://jite.org/documents/Vol3/v3p065-081-127.pdf

Lunce, L. (2004, October). Computer simulations in distance education. *International Journal of Instructional Technology and Distance Education, 1* (10)

Martell, K., & Calderon, T. (2005). Assessment of student learning in business schools: What it is, where we are, and where we need to go next. In K. Martell & T. Calderon, *Assessment of student learning in business schools: Best practices each step of the way* (Vol. 1, No. 1, pp. 1-22). Tallahassee, Florida: Association for Institutional Research.

Mitchell, B. C. (2006). Using networked digital simulations in business courses. *Proceedings of the UMES Office of Instructional Technology 2006 E-Learning Seminar.* July 11, 2006. Princess Anne: Maryland

National Center for Postsecondary Education. (2001) *Student assessment in higher education: A comparative study of seven institutions.* University of Michigan.

Orlich, Harder, Callahan and Gibson. (2004) Teaching Strategies: A Guide to Better Instruction. New York: Houghton Mifflin.

Old Dominion University. (2006). The history of assessment at Old Dominion University. Retrieved 3/13/06 from http://www.odu.edu/webroot/orgs/ao/assessment.nsf/pages/history_page

Page. D.(2006). 25 tools, technologies, and best practices. *T. H. E. Journal, 33* (8). Retrieved from http://thejournal.com/articles/18042

Petrides, L. (2006). Data use and school reform. *T. H. E. Journal, 33* (8). Retrieved 3/13/06 from http://thejournal.com/articles/18041

Petkov, D. and Petkova, O. (2006). Development of scoring rubrics for IS projects as an assessment tool. *Issues in Informing Science and Information Technology Education. 3*, 499-510). Available at http://informingscience.org/proceedings/InSITE2006/IISITPetk214.pdf

Pellegrino, J., Chudowsky, N., & Glaser, R. (2001). *Knowing what students know. The science and design of educational assessment.* Washington, DC: National Academy Press.

Pearson, D., Vyas, S., Sensale, LM, & Kim, Y. (2001). Making our way through the assessment and accountability maze: Where do we go now? *The Clearing House, 74* (4), 175-191.

Popper, E. (2005). Learning goals: The foundation of curriculum development and assessment. In K. Martell & T. Calderon,

Assessment of student learning in business schools: Best practices each step of the way (Vol. 1, No. 2, pp. 1-23). Tallahassee, Florida: Association for Institutional Research.

Ridgway, J., McCusker, S., & Pead, D. (2004). *Literature Review of e-Assessment.* Bristol, UK: Nesta Future Lab.

Schneider, C. G. (2002). Can Value Added Assessment Raise the Level of Student Accomplishment? Peer Review, Winter/Spring 2002. Retrieved from http://www.aacu.org/peerreview/pr-sp02/pr-sp02feature1.cfm

Swearington, R. (n.d.). A primer: Diagnostic, formative, & summative assessment. Retrieved 2/4/06 from http://www.mmrwsjr.com/assessment.htm

Urciuoli, B. (2005). The language of higher education assessment: Legislative concerns in a global context. *Indiana Journal of Global Legal Studies, 12* (1), 183-204.

Vendlinski, T., & Stevens, R. (2002). Assessing student problem-solving skills with complex computer based tasks. *Journal of Technology, Learning, and Assessment, 1* (3). Available at http://escholarship.bc.edu/jtla/vol1/3]

Walcott, S. (2005) Assessment of critical thinking. In K. Martell & T. Calderon, *Assessment of student learning in business schools: Best practices each step of the way* (Vol. 1, No. 1, pp. 130-155). Tallahassee, Florida: Association for Institutional Research.

Walvoord, B. E., & Anderson, V. J. (1998). *Effective grading: A tool for learning and assessment.* San Francisco: Jossey-Bass

Wiggins, G. (1990). The case for authentic assessment. *Practical Assessment Research and Evaluation, 2* (2).

Wright, B. (2004, October 1). An assessment planning primer: Getting started at your own institution. Presentation Made at the *13th Annual Northeast Regional Teaching Workshop.*

Wu, D, & Hiltz, R. (2004, April). Predicting learning from asynchronous online discussions. *Journal of Asynchronous Learning Networks, 8* (2), 139-152.

Note

The original version of this paper was previously published in the Volume 5, 2006 issue of the Journal of Information Technology Education http://www.jite.org.

Authors

Dr. **Nicole A Buzzetto-More** is an assistant professor and assurance of learning coordinator in the Department of Business, Management, and Accounting at the University of Maryland Eastern Shore. Originally from New York, she has a doctorate in Communications and Instructional Technology from Columbia University, a masters in education in Communications and Instructional Technology from Columbia University, a masters of science in Communications from the College of New Rochelle, and a bachelor of arts degree from Marist College. Her intellectual interests include the uses of digital technologies to augment teaching, learning, and assessment. She holds several grants that support her various instructional technology research and development projects. She has published numerous papers in referred journals and is a frequent presenter at conferences across the globe. In 2005 she was honored by the American Distance Education Consortium.

Dr. **Ayodele Julius Alade** is interim dean of the School of Business and Technology at the University of Maryland Eastern Shore. Originally from Nigeria, he has degrees in economics from Cuttington University (Cum Laude) in Liberia and the University of Utah. His areas of specialization include: operations management, quantitative methods, international and developmental economics, industrial economics, and corporate finance. He has over 60 publications and made numerous presentations, and has and continues to hold a number of large-scale grants all of which are interdisciplinary and international in scope. He has been involved and conducted a number of international assignments sponsored by USAID and has received several honors and awards.

Dr. **Dandeson Panda** is an assistant professor of Marketing in the Department of Business, Management, and Accounting at the University of Maryland Eastern Shore. He earned his doctorate from Howard University and his interests include program development, ethics, and international marketing.

Buzzetto-More, N. A. (Ed.). (2007). *Advanced Principles of Effective e-Learning*
Santa Rosa, California: Informing Science Press. (pp. 245-266)

Chapter 10

Process Diaries: Formative and Summative Assessment in e-Learning courses

Peter Bergström & Carina Granberg

Introduction

Ever since information and communication technology made it possible to communicate whenever you like and share written information without waiting for the morning post, higher education has used different e-learning platforms to bridge the physical distance between teachers and students in e-learning courses.

However, even though the platform supports collaboration, communication, and assessment, and while students are active in discussions and virtual seminars, there is a lack of "presence" in the courses. Teachers may feel that they lose contact, not with the students themselves, but with the creative process during the course through which students construct new knowledge.

Education is constructed out of teaching and assessment. These two components should be designed in line with the aims of the course (Gronlund, 2003). The summative assessment at the end of each course is a necessary examination but pays no attention to the important process in which a student's competence grows. This way of evaluating knowledge is focused on the product and not on the student, the very opposite of Knowlton's (2000) view that focusing on the student's processes is a necessary condition for high-quality e-learning. This summative way of looking upon knowledge is not consistent with today's way of evaluating learning processes. There is a need for a method that places the students and their creative processes in focus, a formative assessment that allows the teacher to be present during the process. This presence provides opportunities to create a dialogue and to encourage, influence, and individualize students' learning processes (Airasian, 2005).

For e-learning teachers to have access to students' learning progression, they need an instrument to communicate this process. The use of process diaries to support student-focused methods could be one solution to this challenge. These diaries provide a digital place where teachers can respond to students' work processes, and give feedback to encourage, challenge and guide them. This method of formative assessment could be one way to bridge the distance between teaching and learning in e-learning courses.

The Department of Interactive Media and Learning (IML) at Umeå University has a long tradition of e-learning. Since 2002, process diaries have been used and evaluated at the department to examine whether it is possible to use diaries for formative and summative assessments. Accordingly, process writing has been given special attention in course evaluations. The results and some reflections on these questions are analyzed later in this chapter.

Conception

E-Learning and Assessment

Garrison and Anderson (2003) raise the question of presence in e-learning settings. From their point of view, presence could be described as social, cognitive, and teaching presence. E-learning requires a refocus and repurposing of assessment strategy (MacDonald, Weller, & Matson, 2002). From that perspective Garrison and Andersson argue that e-learning activities must be integrated in the course. Formative assessment is highlighted in work with the learning processes (Garrison & Anderson, 2003; Hudson, Hudson, & Woldinga, 2005). Hudson, Hudson and Woldinga argue that the teacher or tutor plays an important role to support students' processes through dialogue with them. The framework of process-oriented assessment includes flexibility, reflective writing, dialogue, collaboration, and feedback (Garrison & Anderson, 2003; Hudson, Hudson, & Woldinga, 2005; MacDonald, Weller, & Matson, 2002). Therefore, process diaries used to support process-oriented assessment may strengthen that presence and support in e-learning courses.

Summative Assessment

The purpose of summative assessment is to obtain an overview of previous learning and measure to what extent the students have reached the learning goals in the course syllabus (Black, 1998). Summative assessment is associated with consequences (Gronlund, 2003). An example is a high-stake test (Roos, 2002) such that a student who passes the test can continue to the next level of learning. Otherwise, the student is not allowed to continue. Roos argues that there could be social effects from a high-stake examination. The result of the assessment is primarily used for the grading the student but may also be used to evaluate the effectiveness of instruction rather then support learning (Gronlund, 2003). Gipps (2005) takes a strong position, highlighting the widespread concern about only using summative assessment for grading and arguing that grades or marks do not necessarily enhance learning.

Formative Assessment

Black (1998), Black and Wiliam (1998), and Torrance and Pryor (1998) describe a context of formative assessment in primary or secondary school. The concept of formative assessment and e-learning is described in a number of studies (e.g., Garrison & Anderson, 2003; Hudson, Hudson, & Woldinga, 2005)

The purposes of formative assessment are seen as twofold (Black, 1998). Firstly, it gives information to the teacher information on what could be done to support students in each of the course goals. Secondly, it is a tool to define what students should focus on to reach these goals. According to Gronlund (2003), formative assessment is used to measure a student's outcome in a specific area according to the goals of the course. The purpose of formative assessment is not to grade students but to support their learning and determine how to remedy any learning deficiencies. Black (1998) argues that all formative assessment is to a degree diagnostic. One interpretation of formative assessment is expressed as "encompassing all those activities undertaken by teachers and/or by their students, which provide information to be used as feedback to modify the teaching and learning activities in which they are engaged" (Black & Wiliam, 1998, p. 2). This implies that feedback is important in the process of formative assessment. Gipps (2005) explains that feedback can improve student learning. Ramaprasad (1983) argues that feedback must be connected to some kind of reference point. Feedback information identifies the student's knowledge level in relation to the goals in the curriculum for the specific task or subject. Formative assessment is a low-stake assessment. The fact that low-stake assessment does not have social consequences for the student is important (Roos, 2005). Roos argues that this could have positive effects for students when the assessment outcome is used for guidance.

Torrance and Pryor (1998) have identified two conceptually distinctive approaches to classroom assessment which they describe as convergent and divergent formative assessment. *Convergent assessment* implies that learners will follow a predefined pattern or progression that the teacher has decided. The practical implication is that the teacher asks questions that he/she already knows an answer to and the teacher awards credit when the learner can give the right answer. There is a lack of dialogue or collaboration between the teacher and the learners. This view of learning is behavioristic in nature.

In *divergent assessment*, the focus is on the learner and the learning process. The teacher and learners work with open questions and importance is placed on how the students think. This view of learning is constructivist, and the aim is that the teacher teaches in the zone of proximal development (Vygotsky, 1978). This approach of divergent formative assessment is consistent with the approach to process dairies. Black (1998) concludes that surveys on the practice of formative assessment have been done in several countries, but its potential to improve teaching and learning has not yet been exploited; this is also the case in higher education.

Formative and Summative Assessment – Complementary Approaches

Black (1998) explains that there is a problem for teachers who use both formative and summative assessment, arguing that summative assessment can tempt teachers to formalize their assessment work so it is isolated from the development of learning. According to Boud (1995), another problem is that both teachers and students are fostered in a culture of summative assessment that tends to subdue the ideas of formative assessment and learning. Wedman (1988) discusses assessment in a school context. He argues that the results of an assessment are a source of learning and should not be neglected, though teachers often do so. Summative results can be a source for learning and could be used with a formative purpose.

Is it possible, from such a point of view, for teachers to use formative assessments of a task and then judge the student? Black and Wiliam (1998) report that teachers in Britain have tried both summative and formative methods of assessment. There seem to be two groups of teachers: one group argues that teachers could not do both, while the other group argues they could assess using both methods together. Torrance and Pryor (1998) suggest that judgment also can be a part of the formative assessment. In the context of higher education in Sweden, teachers have the opportunity to assess students using both methods, but they must always conduct a summative assessment at the end of a course.

Feedback

Feedback is an important component in formative assessment. According to Weaver (2006) and Greer (2001), feedback should be related to criteria such as helping students to clarify problems, helping students to explain gaps in knowledge and understanding, providing constructive criticism and positive comments to improve self-confidence. All these aspects have the purpose of letting the student know how to improve. With that in mind, feedback on coursework should be valuable also for summative assessment if both formative and summative assessments are combined.

However, teachers must provide guidance for how students will use the feedback. Weaver (2006) states that feedback that is too general or vague results in greater student confusion. Similarly, Greer (2001) confirms that students are not satisfied if the critical evaluation is unclear. It is important not to focus only on the negative because students may drop out of the course if the negative impact is too strong. The language and the way in which we express the feedback are important for how the student understands the message.

Weaver (2006) explains the problem of when students perform well and it is hard for the teacher to find gaps in knowledge; those students tend not to gain any benefit from feedback. From the perspective of the teacher, one obstacle is that it may be difficult to provide every student with quality feedback during the course.

Dialogue

Gorsky and Caspi (2005) perceive dialogue as intrapersonal and interpersonal. *Intrapersonal dialogue* has to do with our own mental processes when we are solving a problem or attending a lecture. *Interpersonal dialogue* deals with both social processes and episodes regarded as part of solving an assignment. Renshaw (2004) argues that dialogue is a process in which we create an intellectual fellowship with one another. However, there is also a view of dialogue as an individual process in which we deal with our reflections and thoughts. Dialogue connected to an assignment is described

> *… as a discursive relationship between two participants characterized by thought-provoking activities, such as hypothesizing, questioning, interpreting, explaining, evaluating, and rethinking issues or problems at hand. (Gorsky & Caspi, 2005, p. 140)*

This conception and definition of dialogue is very close to that used in our work with process diaries.

Pedagogical Approach

When students start to write a process diary, they are required to begin by describing their own view of the course goals. The purpose is to get a view of what the students aim for, how they understand the goals, and what they have most interest in. When students start to work with the assignments, they are supported as outlined in Figure 1.

Figure 1 shows how students worked during and after an assignment, in two stages. Before the assignment starts, students are guided to read the instructions carefully and write down what their own goals are and how these fit with the course goals. Students describe what they will aim to learn during an assignment. Teachers give feedback on how the students' personal goals fit with the course goals.

During the assignment, students document their own learning process and what they are learning in relation to course goals. Teachers give students feedback frequently during their work, and it is therefore important that students write regularly.

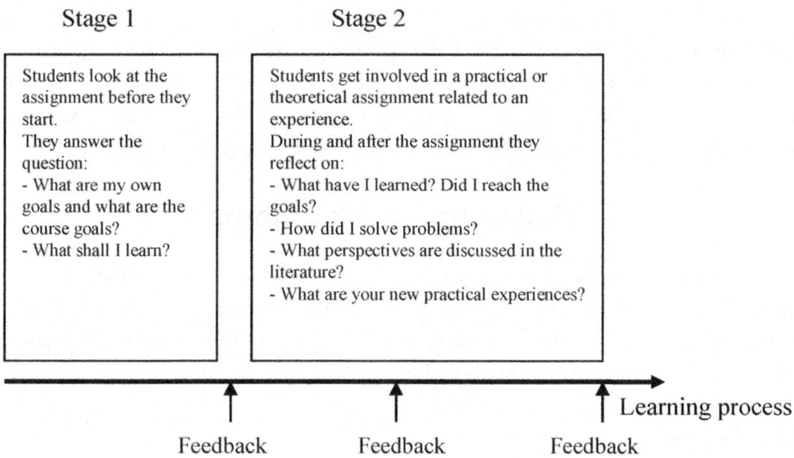

Figure 1. The learning process using process diaries.

At the end of the assignment, students reflect on and analyze what they have learned and whether or not the course goals have been reached. In that context it is important for the students to analyze problems and how they have solved them. In describing the personal learning process, reflection, analysis, argument, and critical thinking about course literature are important tools.

Process Diary or Reflective Journal?

Egidius (2006) explains the word "process" as a term for psychological change where each step provides opportunities for the next. The word "diary" indicates that the activity should be documented frequently, an ongoing process. Rogers and Clare (1994) explain the term as including both reflection and learning process. One important purpose is for learners to take control of their own learning. In the framework presented, goals and learning process are central parts in the pedagogical work.

The literature on pre-service teacher education suggests the term "reflective journal" (Francis, 1995. Spalding & Wilson, 2002. Keppel & Carless, 2006). The format is differently described and the concept tends to focus on student reflection.

However, process diaries are concerned with the whole learning process, in which student reflection is just one part. Process diaries are defined here as a student-driven assessment where knowledge, goals and criteria will be formulated and through which the personal learning process will be visualized and reflected upon in a dialogue with the teacher, the subject, and the student self (Rogers & Clare, 1994).

Pedagogical Arguments

A number of arguments for using process diaries in e-learning have been derived from the foregoing literature review. The arguments are guidance, progress and learning, student perspective, integrating theory and practice, and monitoring and control.

Guidance

In contrast to summative assessment, performed when the students' assignments have been completed, the process diary illuminates the creative learning process and generates opportunities for dialogue and guidance throughout the course. Even if the students do not actually ask for help, their process writing will disclose if they have misunderstood, drawn incorrect conclusions, or are having trouble in getting on with their work. This information gives the teacher opportunity to guide the students in their learning process.

Progress and Learning

Process diaries support the opportunity to acquire information about the quality of the knowledge the students actually obtain throughout their studies. Some students increase their competence in working with the tasks but do not attain the higher marks because of poor initial knowledge. Use of a method focusing on the learning process provides the opportunity to give these students positive feedback. Teachers need formative assessments to perceive and comment on individual progress, a method through which students start by describing their competence, their need for new knowledge, and their aims for knowledge development. This formulated baseline, the dialogue throughout the course, and the final result will together make the knowledge development visible so that credit can be given to hard work and increased knowledge.

Student Perspective

Process diaries can decrease the feeling of loneliness among students enrolled in e-learning courses, where students are left very much alone unless they make contact to receive assistance. Formative assessments provide both students and teachers with a more active role, creating a dialogue about students and their processes of increasing knowledge and proficiency.

During the course, the teacher focuses on the student and not the final product. Students do not need to wait until the final assessment to get feedback, which they may need earlier in the course to avoid mistakes and failures.

Integrating Theory and Practice

The use of process diaries encourage students to show their interpretation and understanding of course literature, lectures, and discussions in an e-learning course in which students display their skills by creating pictures, movies, websites etc. The intellectual process through which theory presented in the literature and practical work integrate with each other can be lost if students are not obliged to illustrate their learning process. A diary in which this process can be put in writing and in which the students can reflect on their work by using course literature as a frame of reference provides a theoretical depth to the assignment, and the teacher will be able to assess the theories beyond the actual product.

Monitoring and Control

The need for monitoring and control relates to formative as well as summative assessment. It makes it possible to monitor students' presence in the course so that their work will lead them to the course aims. Formative assessment furthermore reduces the risk of plagiarizing. Digital communication supported by process writing creates a dialogue about the learning process; this individual contact decreases the difficulties in identifying individuals, so there will be less uncertainty as to who actually handed the project in for the final examination.

A Case Study of Process Diary Usage

This case study examines the use of process diaries in two e-learning courses: "ICT for Education," a three-week course and "ICT and Learning," a ten-week course.

The diaries used in these courses consisted of digital, shared documents in a Learning Management System (LMS) that provided students and teachers with the opportunity to read and write on a continuing basis.

Whilst the sample size is limited, the data illuminate student and teacher experiences of process writing in these courses. So whilst the study does not provide grounds for generalisation, it has constituted a basis for ongoing course development.

"ICT for Education" Course

The participants in this course were students and teachers in higher education. Data were collected from one assignment in which 25 students (15 women and 10 men) answered a question in which they compared a test with multiple-choice questions and the process diary they wrote during the course. The students were asked to comment on advantages and disadvantages of the two different forms of assessment. The course took place during the fall semester of 2004 and the spring and fall semesters of 2005. The technical framework consisted of editable FirstClass documents in which students wrote their process diaries. Students received feedback every third week during the semester.

Advantages of Process Diaries

Twenty-two students gave positive evaluations of the process diaries. Three different aspects were described as benefits when compared to a multiple-choice test. The students described aspects of learning, feedback, and control.

Twelve students expressed different views of learning. Some students (N=9) expressed the view that the process diary supports a higher order of learning, knowledge, or understanding. The term "reflection" (N=6) was used in the context of deeper learning. The reflection was expressed as both "critical" and "personal". For example:

"*The assessment form supports deeper learning, self-critical reflections.*"

Twelve respondents mentioned feedback as a benefit of the use of process diaries. Feedback is used to communicate with the student about their efforts and what they have written about, towards the course goals. The students view feedback as confirmation from the teacher on their effort. For example:

"*The teacher guides you that you are on 'the right way.'* "

Students also stated that the feedback must be active. When feedback was active there was a dialogue between the teacher and the learner.

Some students (N=8) stated that the process diary supported the learning process over time. They argued that learning is continuous and that they could be assessed in a formative way during their studies. For example:

"*The teacher can follow the development of the student and at an early stage discover if he/she starts to drift apart.*"

Monitoring progress was viewed from two aspects by a smaller group of students (N=5). One aspect of monitoring progress, they said, is that it is harder to cheat because the diary is too personal. The second aspect is that the diary helps teachers to have control of students' progress. For example:

"*The teacher has control of the students, how they think, follow the study pace and solve the assignments.*"

Only two students commented on how the personal goals and course goals worked for them in the context.

Disadvantages and Difficulties with Process Diaries

Students (N=11) highlighted some disadvantages or aspects that could be further developed. Two related to understanding the process diary and feedback. Some students (N=8) reported difficulties in understanding the assessment form. One problem was connected to the term "what" i.e., what students should write, and what is expected of them. For example:

"*I thought from the start that it was a little bit difficult to really know what I should write in it.*"

Another problem was the need for clear instructions with regard to "how" and "why" students write a process diary. One argument is that the diary is too open that the assessment part is difficult for the teacher. Accordingly, process diaries must be well developed before using them. For example:

"*The teacher must carefully think what frames should be used for the process diary.*"

When students write process diary entries they connect their learning to course readings. Students can describe how they are struggling. For example:

"*The disadvantage is that it is sometimes difficult to know if my writing is really relevant. 'I experience…' 'I feel that…' and 'My opinion is that...' are often introductions that not are appreciated without heavy references to course literature. I try to connect much of my arguments toward the literature but there are still long parts without references. I do not know if this is a disadvantage, maybe is it good…?*"

The process diary uses feedback to describe knowledge gaps and areas that could be further developed. In relation to this, students are not clear about when the process diary is really finished. There is a disadvantage if the feedback is late:

"*A disadvantage is that the teacher could have problems with time to conduct a constructive dialogue with all students in the course.*"

"ICT and Learning" Course

In the spring of 2005, fourteen students participated in the "ICT and Learning" course The participants consisted of teachers working in primary and secondary school and student teachers. The course aimed at giving the participants proficiency to work as teachers in e-learning courses. The tasks were theoretical and practical, and the students kept a diary to reflect on their work. The purpose of process writing in this course was to illuminate and support the students' learning processes. Furthermore, the aim was to integrate course literature into the working practice. The process diaries were structured with headlines and questions for reflection, and the teachers gave feedback every week.

The empirical data used to evaluate this course consisted of three (N=3) teacher verbal evaluations, ten (N=10) students´ course evaluations and two (N=2) process diaries. The students filled in the evaluations anonymously and the students who wrote the diaries and have been quoted in this text gave their permission.

Students' Reflections

To evaluate the use of process diaries, the students were asked to grade two questions regarding process diaries and give their comments on their grading. The questions looked into the relevance of the process diary and whether the working method served its purpose as a formative assessment.

Relevance

Seven out of ten students found process diaries very relevant; they expressed the importance of receiving feedback throughout their work. They pointed out that writing and obtaining comments make it possible to get confirmation that they were on the right track, obtain correction of misunderstandings early in the course, and receive assistance such as pointing out and correcting lack of knowledge, etc. One of the students wrote:

"*The idea of process writing and receiving feedback that I can respond to is great. To know that someone is there…*"

Fulfilling its Purpose

Eight out of ten students experienced that working with process writing served its purpose to a satisfactory or very satisfactory level. However five (N=5) students expressed some criticism that continuous feedback is very important if the process writing is to serve its purpose. Two of these students wrote that they lost interest in writing because of the time delay between their writing and the teachers' feedback. On the other hand, just as many (N=5) of the students commented that they felt noticed and received the support they needed to solve technical or theoretical difficulties:

"*I appreciate the teacher's good advice that helped me from getting stuck in the empty nothingness.*"

Some of the students (N=3) had initial problems putting their thoughts and reflections in writing in the diary. These difficulties are discussed in the evaluations. For example:

"*Reading your feedback I understand that there is a 'right way' of writing in the diary, but I do not know how to write the way you want me to.*"

A small group of students (N=2) expressed difficulties in dealing with feedback. Instead of feeling supported, they felt accused, and one of them wrote in the diary after receiving feedback in order to correct some misunderstandings:

"*I felt sorry when I read your feedback, that you failed me so early in this course.*"

The difference between formative and summative assessments is obviously not clear to this student.

Teachers' Reflections

After the course examinations, a meeting was held in which three of the participating teachers discussed their experience, progress, and setbacks.

Process writing

The teachers argued that formative assessment made it possible to individualize. However, it is important to make sure that this individualization does not lead the students away from the aim of the course but takes them beyond it.

Teachers described how the diaries have potential to examine the course literature and give the practical exercises a theoretical background. The theoretical and practical assessments will in this way be integrated. Students who are having problems in relating their practical work to course literature can be given guidance early in the course and teachers can assist them to use the theories to support their ongoing work.

All teachers commented that they felt that students were not used to this kind of formative assessment and it took time for them to displace some of their focus from the final product to the ongoing working process. Keeping a process diary made demands on the students' capability of self-assessment, and some of the students needed a lot of support to be able to describe their own knowledge and increasing understanding.

The teachers argued that an early start in writing in the process diary is vital to fulfill its purpose of formative assessment. It is important to be able to clarify poorly formulated instructions and misunderstandings early in the course as well as the purpose in keeping a diary.

Giving and receiving feedback

Reading the students' diaries and giving feedback is a way of bridging the physical distance in e-learning. Teachers expressed the view that these interactions may to some extent compensate for the personal contact you can get in the physical classroom.

Giving adequate feedback is of great importance and has to be constructive, individually suitable, and timing. Teachers, however, pointed out that there is a risk that individual guidance may give the students the answers instead of supporting them in researching answers for themselves; it is important for teachers to clarify that they will not provide a shortcut to higher marks by this weekly guidance. Some students expressed this frustration in their diaries:

"*Why can you not be more specific in your feedback about what you want me to write to get higher marks?*"

The time delay in feedback is one of the problems mentioned by the teachers. Students would prefer to have immediate comments on their reflections and to get quick answers to their questions. That is of course not possible when teachers are engaged in and have to spread their time across several courses. Teachers stated that it is of great importance to clarify to the students how much and what kind of feedback they can expect as well as what the response time will be.

Analysis

Is it possible to use process diaries as an alternative assessment where formative and summative assessments are integrated? Do the empirical results from these evaluations support this approach to integrate assessments, and to what extent does this method correspond to stated arguments? This analysis will look into these questions.

Formative Assessment

Formative assessment is aimed at supporting learning towards the course goals (Black, 1998; Gronlund, 2003; Torrance & Prayor, 1998). According to student responses, the argument related to using process diaries for *guidance* is clearly fulfilled and recurs as an important instrument to support some of the other arguments. In order to give students guidance, feedback is of great importance (Greer, 2001; Weaver,

2006). Students express this guidance as positive. Results show that the feedback is perceived as support, encouragement, and confirmation that they are working towards the aim of the course.

Feedback is important in relation to *progress and learning*. Students who participated in the "ICT for Education" course in particular commented that they found this guidance helpful towards receiving a deeper understanding, for promoting self-reflection, and to move the focus from the marks given at the end of the course to their own learning process. Ramaprasad (1983) highlights the importance of illuminating the students' level of knowledge in relation to curricular aims. Without this baseline, teachers note that they will have no opportunity to observe and to comment on students' learning progress. The teachers' main issue is to support learning towards the aims of the course by giving feedback, but students have difficulty in describing their work towards goals. Most students (80%) strongly agreed in this study that the process diary is relevant in order to accomplish the assignments, but there were few responses in which the goals and feedback were seen as connected to each other.

Feedback is also seen to initiate interactions in terms of being "noticed" and "creating a dialogue." These are the key issues in relation to the *student perspective*. The intellectual fellowship that Renshaw (2004) describes becomes visible when the student declares the process as a three-part dialogue between the teacher, subject and the student's self. These interpersonal and intrapersonal dialogues (Gorsky & Caspi, 2005) not only support the learning process but are important parts of the social process which can prevent the student from feeling "left alone."

Summative Assessment

Swedish teachers who have to grade their students are obliged to perform a summative assessment. According to Black (1998) and Gronlund (2003), the only function of summative assessments is control and to measure to what extent the students have reached the course goals. Therefore, the argument for using process diaries in order to support summative assessment is to control, and to some extent, integrate theory and practice.

At the end of each course in this study, the process diaries were handed in to become a part of the final assessment. The students made few comments on the summative use of the diaries besides some expression of "being afraid to not be able to write the way the teacher wants."

The teachers in the study argued that process writing encouraged the student to reflect on how they used course literature to solve their assignments. This way of *integrating theory and practice,* supports formative assessment as a means of helping students to understand the literature, as well as summative assessment where the students' understanding of literature can be assessed.

The argument of *monitoring and control* was discussed by some students in this study, who concluded that process writing would make cheating more difficult. The summative assessment itself constitutes a control of the students' achievements. This is the high-stake measurement in which students who do not reach the aims by the end of the course can fail (Roos, 2005). Gipps (2005) argues that final assessments do not necessarily stimulate the learning process. However, students in the study stated that they would increase their efforts if they knew that in the final assessment they might be given higher marks (e.g., pass with distinction).

Integrating Formative and Summative Assessment

Black (1998) views summative and formative assessment as end points on a spectrum. They are so different that they could not be combined. However, through this approach with process diaries the purposes of both summative and formative assessment are reached. Gronlund (2003) and Black (1998) argue that course goals can be used as a reference for guidance as well as the final assessment. The aim of formative assessment is reached during the course when students get feedback and teachers support their learning process towards goals. The summative purpose is reached when teachers at the end of the course make a judgment on the final process diary to what extent the goals have been reached. Accordingly, the summative–formative spectrum is still present, but the format of the process diary integrates both summative and formative purposes.

Discussion

When integrating summative and formative assessment an urgent issue is that students must understand how they are being assessed. Students state their problems in terms of "how," "what," and "why" questions.

In the case of "how," some students do not understand that the process diary is formative and feedback will be interpretive. Students need to understand that they not will be failed early in the course. The "what," could first be explained by providing students with instruction about what they will describe as having been learned. They need this because they do not have experience in writing about their own learning. Second, students ask what they should be writing to get a higher grade. This could indicate that students are not used to working towards goals and criteria or that teachers are not familiar with explaining these goals and criteria carefully enough. "Why" is connected both to the issue of monitoring and control and to the purpose of having contact with students during the course so as to be able to give them individual support.

Although alternative assessment may fulfil its purpose, empirical data point out some of the difficulties in using process diaries. However, the most frequent criticisms in the empirical data of this study are in the students' experience with the time delay between their writing and teachers' feedback. This problem is discussed by Weaver (2006); teachers respond to this criticism by pointing out the importance of clarifying in advance how feedback will be provided in order to give the students reasonable expectations of this dialogue.

Summary and Future Work

Boud (1995) argues that students are more familiar with summative than formative assessments. This lack of practice with the formative method becomes apparent when students state that they do not find "the right way of writing" and do not comprehend the difference between a final exam that often requires predetermined "right" answers and a formative assessment where there is no obvious "right" or "wrong." When teachers use open questions in order to focus on students' processes (Torrence & Pryor, 1998) some of them initially get frustrated. Dealing with open questions like: "Explain your choice of method and discuss..." is a process itself, in which students will not find any "right" answers in the course literature but have to construct their knowledge to reflect on the given questions.

This unfamiliarity with formative assessment also appears when students interpret teachers' corrections during the process as a statement of failure. The intentions of this formative, low-stake assessment

(Roos, 2005) are confused with the summative assessment, and the student misses the guidance offered. But even if there is no such thing as failing the formative assessment, there are students who have difficulty in receiving a negative critique and have a hard time using this feedback to work their way through the process. They feel that they have failed and as Weaver (2006) notes, when students receive mainly negative feedback they become "demoralized and angry"—reactions that may hold back the learning process.

The approach used for process diaries (see Figure 1) indicates that a focus on course goals is not of great importance to students. Our results show that it is urgent that students focus on course goals in their learning process. The study identified aspects of formative assessment and process diaries that could be further developed. In Figure 2, the approach has been reworked in order to focus more on course goals at Stages 1 and 3.

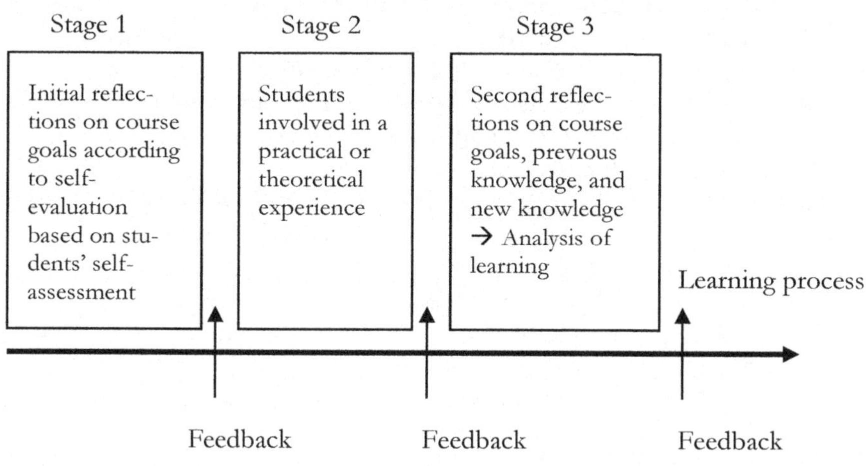

Figure 2. Learning process using process diaries focusing on course goals

Figure 2 shows a future model for working with process diaries during the learning process. In Stage 1, students reflect on their knowledge in the area to be studied. The teacher provides feedback on the prior knowledge and identifies gaps to focus on. In Stage 2, students get a theoretical or practical experience of the subject that is the focus of the assignment. Students pass to Stage 3 and subsequently construct a new analysis of the knowledge and learning outcomes according to curriculum, course goals, and previous knowledge. Teachers provide feedback with the purpose of identifying knowledge gaps and participating in a dialogue around the learning outcomes. If knowledge gaps are identified, students must continue the process to bridge those gaps, with teachers providing new feedback.

Process diaries, integrating summative and formative assessment focusing on the course goals, create the opportunity to bridge the distance between the teacher and students' learning process as well as the gap between students' initial (present) level of knowledge and the aims of the course.

References

Airasian, Peter W. (2005). *Classroom assessments: Concepts & application.* 4th edition. Boston: McGraw-Hill.

Black, P. (1998). *Testing: Friend or foe? Theory and practice of assessment and testing.* Philadelphia: Open University Press.

Black, P., & Wiliam, D. (1998). Assessment and classroom learning. *Assessment in Education: Principles, Policy & Practice 5*(1), 1-74.

Boud, D. (1995). Assessment and learning: contradictory or complementary? In P. Night (Ed.) *Assessment for learning in higher education.* London: Kogan Page.

Egidius, H. (2006). *Termlexikon i pedagogik, skola och utbildning.* [Dictionary of terms in pedagogy, school and education]. Stockholm: Studentlitteratur.

Francis, D. (1995). The reflective journal: a window to preservice teachers' practical knowledge. *Teaching and Teacher Education, 11*(3). 229-241.

Garrison, D.R., & Anderson, T. (2003). *E-Learning in the 21st Century: A framework for research and practice.* New York: RoutledgeFamler.

Gipps, C. (2005). What is the role for ICT-based assessment in universities? *Studies in Higher Education, 30*(2), 171-180.

Gorsky, P., & Caspi, A. (2005). Dialogue: a theoretical framework for distance education instructional systems. *British Journal of Educational Technology, 36*(2), 137-144.

Greer, L. (2001). Does changing the method of assessment of a module improve the performance of a student? *Assessment & Evaluation in Higher Education, 26*(2), 127-138.

Gronlund, N. E. (2003). *Assessment of student achievement.* Boston: Allyn & Bacon.

Hudson, B., Hudson, A., & Wolding, S. (2005). Supporting learners in an international online learning community. In J. Fragner (Ed.) *eLearning in Higher Education*. Linz: Institute of Comparative Education.

Keppel, M., & Carless, D. (2006). Learning-oriented assessment: A technology-based case study. *Assessment in Education, 13*(2), 179-191.

Knowlton, Dave, S. (2000). A theoretical framework for the online classroom: A defense and delineation of a student-centered pedagogy. In R. E. Weiss, D. S. Knowlton, & B. W. Speck (Eds.), *Principles of effective teaching in the online classroom.* New Directions for Teaching and Learning *84*. San Francisco: Jossey-Bass.

MacDonald, J., Weller, M., & Mason, R. (2002). Meeting the assessment demands of networked courses. *International Journal on E-Learning.* January-March, pp. 9-18.

Ramaprasad, A. (1983). On the definition of feedback. *Behavioral Science, 28*, 4-13.

Renshaw, P. & van der Linden, J. (Eds.). (2004). *Dialogic learning – Shifting perspectives to learning, instruction, and teaching.* Dordrecht: Kluwert Academic Publishers.

Rogers, M., & Clare, D. (1994). *The process diary: developing capability within national curriculum design and technology – some initial findings.* Loughborough: Loughborough University of Technology.

Roos, B. (2002). *ICT, Assessment and the Learning Society.* European Conference on Educational Research, Lisbon.

Roos, B. (2005). Examination och det lärande samhället [assessment and the learning society]. *Journal of Research in Teacher Education, 12*(1), 155-175.

Spalding, E., & Wilson, A. (2002). Demystifying reflection: A study of pedagogical strategies that encourage reflective journal writing. *Teachers College Record, 104*(7), 1393-1421.

Torrance, H., & Pryor, J. (1998) *Investigating formative assessment: Teaching, learning and assessment in the classroom.* Buckingham, UK and Philadelphia, PA: Open University Press.

Weaver, M. (2006). Do students value feedback? Student perception of tutors' written feedback. *Assessment and Evaluation in Higher Education, 31*(3), 379-394.

Wedman, I. (1988) *Prov och provkonstruktion* [Assessment and construction of assessment]. Stockholm: Utbildningsförlaget.

Vygotsky, L. (1978). *Mind in society: The development of higher psychological processes.*(trans.).M. Cole, V. John-Steiner, S. Scribner, & E. Souberman (Eds.). Cambridge, MA: Harvard University Press.

Authors

Peter Bergström is a lecturer in the Department of Interactive Media and Learning at Umeå University in Sweden. He has a Bachelor of Education for the Compulsory School in the subjects of Math, Science and ICT and learning in classes 4 to 9. He has a Masters degree in Teaching, Research and Assessment of Quality. From 1997 to 2004 he was a teacher trainer for teachers in school and student teachers in ICT for learning. Since 2001 he has worked in a variety of projects that aim to develop e-learning in higher education.

Carina Granberg is a teacher educator and since 2006 a doctoral student at the Department of Interactive Media and Learning at Umeå University in Sweden. She has experience teaching in a variety of on-line courses for in-service teachers as well as student teachers. Her research interests relate to questions about how ICT can support learning processes as well as implementation processes of ICT-supported methods in the context of education.

Buzzetto-More, N. A. (Ed.). (2007). *Advanced Principles of Effective e-Learning*
Santa Rosa, California: Informing Science Press. (pp. 267-292)

Chapter 11

e-Portfolios in e-Learning

Trudi Cooper and Terence Love

Introduction

This chapter presents an approach to e-portfolios that integrates learning, assessment and education administration. The chapter describes some key factors that must be considered by educators and administrators when using e-portfolio systems to maximize educational and administrative benefits to all stakeholders.

E-portfolios offer significant benefits to learners, educators and administrators of education organizations across different types of educational context. For example, e-portfolios can be used to:

- Integrate documentation of learning from both formal and informal educational environments
- Demonstrate students' learning progress
- Demonstrate learners' professional and ethical values
- Provide a repository of evidence which the learner can access to provide support for employment applications, employment promotion, for application to formal courses and to claim exception from formal studies because of prior learning
- Document learners acquisition of professional skills
- Document summative achievements at the end of secondary and higher education
- Document acquisition of graduate attributes (See, for example, the Dearing Report on Higher Education in the UK that re-

garded the graduate 'skills and attributes agenda' as central to higher education http://www.leeds.ac.uk/educol/ncihe/).

- Provide students with personal electronic access to learning, assessment and administration processes.

E-portfolios are an important aspect of e-learning because portfolio-based education provides a coherent framework that can encompass a variety of learning and education approaches and can overcome weaknesses of other assessment systems. Additional benefits accrue when the pedagogic integrity of e-portfolio systems is combined with an online environment to streamline administrative tasks (Love & Cooper, 2004). Future development of e-portfolio systems based on Semantic Web technologies offer opportunities to more comprehensively integrate e-portfolios with 'real world' elements of learning such as human educators and students, live communications, books, 'real world' libraries etc. This integration of the digital and real world elements of learning accessed via the digital milieu is emerging as an important direction for the future of e-learning (see, for example, Wiley (2001), Harman and Koohang (2005), New Media Consortium and Macromedia (Johnson, 2003), and the IEEE definition of learning objects that include non-digital elements that contribute to learning (Hodgkins, 2002)).

New Semantic Web technologies extend the current World Wide Web to offer two significant advantages for e-portfolios and e-Education. It supports the development of e-portfolio systems in which a single highly-portable e-portfolio can be used by an individual to document *all* their lifelong learning, formal and informal, across education programs including self-education and autodidactic learning,, workplaces and educational institutions worldwide. The new Semantic Web technologies also help address issues of interoperability, associated with the current widespread use of XML and similar markup-based methods used in standardization of e-portfolios. This is an important issue. A section of this chapter will explain why XML-based e-portfolio systems are intrinsically problematic because of their limitations in how they control system variety and how new Semantic Web protocols such as RDF and OWL can resolve these problems. In addition, these Semantic Web technologies offer extra e-learning benefits. For example, RDF-based CC/PP protocols facilitate the application in e-learning systems of highly heterogeneous mobile devices (mobile phones, personal information managers etc.) emerging as central to the involvement of young people with e-learning (Anderson & Blackwood, 2004; Manola & Miller, 2004).

The e-portfolio landscape is viewed in this chapter through the four lenses of pedagogy, constituent orientation analysis, systems analysis, and future Web technologies. The chapter has eight parts. The next section describes how key terms and concepts are used in this chapter and discusses their role in relation to their contexts, pedagogical application, equity issues and administrative considerations. The third section provides an overview of e-portfolios. It describes key pedagogic and administrative considerations for the design of e-portfolio systems, and draws attention to the important issue of interoperability. The fourth section outlines the perspectives on e-portfolios of four key stakeholders or constituencies and demonstrates how these perspectives shape priorities in the design of e-portfolio systems. Section five provides an overview of the five main evolutionary stages of e-portfolio development. It locates contemporary 'cutting-edge' developments in e-portfolios at the transition between generation four and five in the developmental process. Section six outlines how changes in online technology, particularly the use of Semantic Web frameworks involving RDF, offer new challenges and opportunities for the development of e-portfolio systems. Section seven explores the implications of integration of e-portfolio systems with other education organization management systems. The conclusion, section eight, summarizes the chapter and identifies opportunities, challenges and future directions in the design and implementation of e-portfolios for e-learning, assessment and education administration. This inclusive approach to e-portfolios is also echoed in the authors' work on learning objects (see, for example, Love and Cooper, 2006).

E-Portfolios: Terms and Concepts

In the literature on e-learning there is inconsistency in the usage of key terms that problematically combines with confusion and lack of clarity about the underlying pedagogical assumptions of those who adopt portfolios for assessment purposes. In this section we define and discuss how particular terms have been used in this chapter and why we avoid some terms commonly found in e-learning discussion of portfolios. We believe it is important to clarify the e-portfolio terminology because clarity in definition can help distinguish between concepts and beneficially shape e-learning discourse about e-portfolios.

Portfolio: An organized collection of documents or artifacts that can be used to demonstrate knowledge, skills, values and achievements, which contains a commentary or exegesis to explain the relevance, credibility and coherence of each artifact or document, and where necessary provides information about standards of performance.

Profile: An extended curriculum vitae or 'a concise biographical sketch' (Mish, 2003, p. 991). A profile contains formal certificates, but would not contain samples of work, and would not usually contain an exegesis because the meaning of the inclusions would be self-evident. A profile is not a portfolio.

E-portfolio: a generic term that includes any form of online portfolio, Web-based portfolio or any portfolio stored or communicated using electronic technologies. For example,

"An e-portfolio is a Web-based information management system that uses electronic media and services. The learner builds and maintains a digital repository of artifacts, which they can use to demonstrate competence and reflect on their learning." (ePortfolio Portal, 2004, http://www.deskootenays.ca/wilton/eportfolios/whatitis.php) The term 'e-portfolio' includes online portfolios.

Discussion

In this chapter, we use 'portfolio' to denote an *organized* collection of documents or artifacts. We distinguish between simple and complex-portfolios. Simple portfolios contain and present 'collections of work' without a commentary or exegesis to interpret the meaning of the collection. Complex portfolios add interpretation, commentary or exegesis that increases the credibility of the portfolio and extends its potential use. Examples of simple portfolios include traditional artists' and architects' portfolios of drawings that do not contain a detailed commentary. When simple portfolios are assessed, the reader must use their judgment to infer the learner/practitioner's skills and formative development from the stand-alone items of evidence found in the portfolio.

Both simple and complex portfolios can be used 'formatively' to demonstrate progress, or 'summatively' to demonstrate achievement. This gives four categories of portfolio

- Formative simple: a 'portfolio as a collection of samples of work showing learning progression'

- Summative simple: a 'portfolio as a collection of best samples of work'
- Formative complex : a 'portfolio as a document that demonstrates progression in learning'
- Summative complex: a 'portfolio as a document that demonstrates skills, values and competencies'

A formative portfolio includes samples of a student's work to demonstrate progression. In contrast, a summative portfolio includes only the best examples of a student's work to indicate the best that the student can achieve.

The simple portfolio is adequate for purposes in which knowledge and skills are capable of self-evident demonstration from a collection of samples of work. Complex portfolios that include interpretation, commentary or exegesis are required when it is necessary to demonstrate skills or achievements that are not totally and completely self-evident from samples presented without explanation. Professional portfolios typically must be complex because they require exegesis to provide satisfactory evidence for the 'higher professional skills' that are not self evident from work samples.

In this chapter the term 'e-portfolio' is used to describe any type of portfolio that is electronically encoded. We avoid the terms 'Web-based portfolios' and 'Webfolios', 'electronic portfolio' and 'online portfolio for two reasons. First, there is contradiction in current usage. For example, D. Love, McKean, & Gathercoal (2004) use 'e-portfolio' to refer to an electronic copy of a paper-based portfolio using Word or a similar word-processing application and 'Webfolio' for a fully digitized portfolio, whereas Batson (2002) uses each term with the opposite meaning. Second, the difference in nomenclature is based upon a single technical feature, 'being "online"', that blurs other important technological differences. For example, it is technically and pedagogically significant to distinguish between whether an e-portfolio is 'transmitted' online or 'stored' online, or both because these locate the e-portfolio system in relation to the different levels of the Open Systems Interconnection (OSI) Reference Model (for an introduction to the OSI model see http://en.wikipedia.org/wiki/OSI_model). From experience, it is more helpful in a time of high technological change to use extra to describe the exact differences between e-portfolio systems characteristics rather than rely on loosely defined terminology.

E-Portfolios: An Overview

Portfolios are currently successfully used across a variety of modes of education in primary, secondary, tertiary and doctoral education as well as continued professional development, self-education, vocational education, and in-service training. Portfolios-based assessment systems can overcome weaknesses of other assessment processes, when they are implemented with proper attention to the purposes of the educational program.

In a critical review of e-portfolios we undertook in 2003, we reported in that most of the e-portfolio implementations we reviewed appeared to fall short of their potential (Love and Cooper, 2004). In many cases, they were inferior to conventional paper-based portfolio assessment. Our review also found widespread neglect of factors necessary to ensure pedagogic integrity and to maximise value for stakeholders. Typically, we found that the design processes for e-portfolio systems overemphasized low-level technical issues to facilitate technical implementation to the neglect of pedagogic goals and processes.

This critical review identified two different kinds of e-portfolio problem. First, some e-portfolios were not what they appeared. Many e-portfolios comprised single pieces of work such as essays or term papers, that had been simply 're-badged' as 'e-portfolios'. This appeared to be being done without any other pedagogic changes other than some support for students to improve their computer presentation skills. We found this representation of single items of work as e-portfolios was widely endorsed including by e-portfolio experts such as Helen Barrett (see, for example, http://electronicportfolios.com/ALI/samples.html). Second, some made attempts to replicate the pedagogic richness of hard-copy portfolios, replacing hard-copy documents with electronic documents. When the pedagogy of the original paper-based portfolio system was well addressed, this approach appeared practically effective although it did not draw on the larger benefits made available by the use of online technologies. The transition into electronic contexts however, raised new philosophical concerns over the management of ethical issues in cyberspace, over new problems of plagiarism and quality control, and raised practical issues relating to managing the electronic files with the related problems of access and security. In addition, we found significant unexploited potential for enhanced management of e-portfolio systems in terms of education administration.

In an earlier publication, we identified that greatly improved stakeholder value could be created through a different approach to the design of e-portfolio systems (Love & Cooper, 2006). We suggested pedagogic and administrative concerns represent the central functional issues in the design of e-portfolio systems, and that technical issues should be regarded as subsidiary to pedagogic and administrative goals. We also found that e-portfolios offered many extra benefits such as automation of administrative functions, improved assignment security, and opportunities for easier detection of fraud and plagiarism. We found, however, that these opportunities were rarely realised in the earlier stages of e-portfolio development. Later sections of the chapter will examine how things have changed since then, and how new technologies such as the Semantic Web and RDF offer an easier path to these benefits.

Portfolios and Pedagogy

Educational trends in the last two decades have changed from an 'inputs' focus of 'what things individuals can be taught' to an 'outputs' focus of 'how and what people have learned'. In parallel, there has been a move away from didactic styles of teaching to constructivist teaching methods. The move towards constructivist approaches to education has had implications for assessment and has led to a growth in the use of portfolios for assessment purposes. Constructivist pedagogies recognize that people learn from sources other than formal education. The 'outputs' focus of contemporary pedagogies means that there has been a renewed interest in 'authentic' learning, 'experiential' learning, debate about the applicability of 'competency based' education in and across different disciplinary fields, and a search for generic 'graduate attributes'. This section outlines the main pedagogical purposes for portfolios within output-focused curricula. We begin with 'authentic learning' and 'experiential' learning because portfolios are popular as an assessment method for constructivist pedagogies that use 'authentic' and 'experiential' learning sometimes in conjunction with competency-based education.

Authentic learning begins from the assumption that learning is more meaningful to students if it is related to the 'real-life experiences' which are relevant to the target group of learners. It also assumes that students will be more motivated to learn if activities are meaningful to them, and that they will integrate and process their learning more fully. The curriculum development process for authentic learning begins with identification of what learning the student group will perceive as relevant in the context of their present and future life experiences. Learning activities are situated in simulated or 'real life' contexts. Authentic learning makes use of situated case studies or scenarios that present 'problems' and issues encountered in 'real life' with the requirement that students use multiple methods available in life situations such as collaboration, dialogue with expert sources, and research to resolve the problem. Through discussion students are encouraged to generalize their understanding to other related contexts. In authentic learning scenarios, portfolios are sometimes used as student-maintained 'workbooks' for formative and summative assessment purposes. Students are encouraged to develop complex portfolios that include reflection on their learning and exegesis to explain and justify the contents of their portfolios.

Experiential learning is 'learning by doing' and contrasts with didactic 'learning by telling'. In experiential learning, assessment tests how well students can apply their learning *in context*. This contrasts with didactic learning, where assessment is used to test how well students have processed the material delivered in the course. Practicum and workplace-based experience is sometimes used to assess the application of experiential learning in authentic or 'real life' contexts. Portfolios are frequently used to assess this learning. Experiential and authentic learning are sometimes allied to competency-based education although this connection is not inevitable.

The focus on educational outputs rather than inputs has led some courses to embrace competency-based education, and has led several countries to identify and list 'generic skills' that *all* graduates should possess. Both these educational movements have looked to portfolios as a suitable assessment or presentation format because portfolios contain 'outputs'.

Competency-based education is an outcomes-based instructional system where the curriculum specifies 'operational' performance measures for key knowledge, skills and values. In competency-based education, assessment is typically by demonstration and in some fields, assessment is purely observational. Competency-based education is least problematic for manual skills, but becomes more problematic for complex 'higher professional skills', where performance is more difficult to judge operationally, and where concepts of 'good performance' are contested (Davies and Durkin, 1991). Portfolios are typically used within competency-based education to collate examples of 'evidence' where an expert has witnessed satisfactory performance. This, however, is problematic in many professional areas where performance standards are contested.

Lists of generic graduate skills or graduate attributes have been developed in several countries, including the UK and Australia, as an attempt to indicate cross-comparability of standards between disciplines. The Australian Technical Network University Group in Australia defines **Graduate Attributes** as 'the qualities, skills and understandings a university community agrees its students should develop during their time with the institution. These attributes include, but go beyond, the disciplinary expertise or technical knowledge that has traditionally formed the core of most university courses. They are qualities that also prepare graduates as agents for social good in an unknown future.' (Bowden, Hart, King, Trigwell, & Watts, n.d). Portfolios are used with Graduate Attributes as a presentation format, rather than as an assessment method, to provide an organized repository of material that can be used to support future job application. Students are encouraged to present samples of their work in ways that demonstrate the key attributes required of graduates.

A related movement has promoted the idea that people are (or ought to be) 'life-long learners'. This concept has become more prominent because post-industrial societal changes mean that most adults will have multiple 'careers' during their working life, each of which will require that they learn and apply different sets of knowledge and skills. Pedagogically, this has stimulated discussion about how students can be supported to plan their own 'life-long learning program'. Related to the idea of lifelong learning are the concepts of 'self-directed learning' and 'independent learning', and the recognition that people learn from life experience, and may be 'self-taught'. All life-long learning approaches to education give learners some control over what, and how, they learn. Current evidence suggests that people will move between many different forms of learning during their lifetime. The problem of how to document prior learning has become an important issue for students, and the corresponding problem of how to credit prior learning has become an issue for formal educational institutions. Many institutions currently ask students to present a portfolio when they apply for formal credit for their prior learning. The portfolio, in the form of a 'lifebook', has been suggested as a future means to integrate documentation of both formal and informal learning.

Lifelong learning is learning directed by the individual and guided by their interests and life goals. Life-long learning might involve formal study, informal learning, autodidactic learning and can be experiential, or a mix of any of these modes of learning. Life-long learning may be either professionally oriented or non-vocational. Lifelong learning can be formally accredited, work-related, or may be completely informal.

Autodidacticism is self-education (Mish, 2004). An **autodidact** is someone who is mostly self-taught, who is self-motivated and has an enthusiasm for self-education. Some people are formally educated in one discipline and self-educate in other disciplines, either as hobbies or in their professional life. The term 'self-taught' is frequently used pejoratively to signify 'non-traditionally educated', and to perhaps imply that learning is less comprehensive, reliable, or less thorough that that acquired by 'formal' from accredited sources.

Self-directed learning (sometimes referred to as **independent learning)** occurs within a formal institution and indicates students have a measure of responsibility for what and how they learn. In self-directed learning, the role of the teacher is to act as a support and resource for the learner. Portfolios have been promoted as a useful means to document all forms of self-education and self-directed learning because portfolios can contain different types of artifact or documentation or other evidence of learning and achievements. A complex portfolio can also include material indicative of standard of achievement of informal learning.

Equity Issues

E-portfolios use computer-based technologies that require learners to have skills in using software. This can result in equity issues. Some groups of students may be unintentionally disadvantaged if they do not have access to the computer technologies, knowledge, support or skills required for e-portfolios. These issues and the financial and access implications of using computer-based technological applications need to be carefully considered before new technological solutions are designed and implemented. Where institutions provide students with access to computer facilities and software automatically as part of their enrolment, or implement a wholly Web-based e-portfolio system to which all enrolled students have access, this goes some way to resolving equity issues, provided there are adequate resources. Some mature-age students will require access to courses and support to help them learn relevant software application skills. School-leavers are more likely to have gained the necessary skills through their schooling.

Administration

The most significant potential gains from the use of e-portfolio systems are in improving the efficiency of education administration to the benefit of students, teachers and administrators (Love & Cooper, 2004). This remains an under-exploited area, although there is now greater awareness of the potential administrative benefits of online course management systems such as Blackboard and WebCT. Currently, using these types of technology, teaching staff are now able to create mark lists and associated statistics automatically, to access reports on students' assessment status, and to communicate electronically more easily with individuals, sub-groups or the whole student cohort. Recently, Blackboard, WebCT and other e-learning system providers have started to include e-portfolio sub-systems. Thus far, e-portfolio provision in e-learning systems has been limited in scope. There focus has been on creating a style of e-portfolio similar to a curriculum vitae or profile. The commentary or exegesis is missing. This is a significant weakness as it precludes the use of these systems for complex portfolios. Practically more important from the point of view of educators, it is likely to result in significantly increased assessment time and workload. The ease with which a student can add any material to their portfolio and the lack of a commentary to explain the evidence means that assessors sift through copious material to find evidence to support their assessment judgments. The pace of change in the e-learning system arena is high. Blackboard and WebCT are about to merge and Blackboard is rumored to have a new e-portfolio system in development for release around 2010.

The capability to administer and manage learning environments is the central purpose of education programs. The use of e-portfolio systems offers the potential of increased sophistication in administration between and across awards, courses institutions and life experiences. This offers the possibility of improvements in learning through the provision of learning environments better aligned to the needs of learners. With suitable online environments, there is scope for integration of e-portfolios into systems that enable educators to map learners' relative development across a range of objectives, to identify future learning, to generate automated feedback reports for future curricula and lesson planning and to receive evaluation from learners about their perceptions of education provision. Developments in some current e-portfolio systems allow students to access parts of their records held by the university and for example, enable students to copy the transcript of their academic achievement to their e-portfolio (Treuer & Jensen, 2003). Alumni can manage their own records through a lifelong possession and control of their e-portfolio. This would offer the potential for learners to, for example, automate production of their CV, course status documentation, or list of certificates. It would also potentially give participating education institutions access to personal data on alumni raising significant data privacy issues.

E-Portfolios stored electronically with a suitable access technology, can potentially simplify administrative processes that support teaching including:

- Automation of administrative functions such as recording of student progress
- Plagiarism detection
- Management of security, avoidance of fraud and improved protection against tampering with students' records
- Quality Assurance reporting

Technical Problems

Interoperability of e-portfolio systems is a significant issue. Interoperability enables students to transfer seamlessly between campuses, courses and institutions, and facilitates integration of formal education, lifelong learning and professional development. In parallel to developments in learning object systems, the preferred approach thus far for larger-scale e-portfolio systems has been to establish a central, secure, complex repository of certification and evidence of competence for each student (see, for example, National Learning Infrastructure Initiative, 2003). Initial attempts to 'electronify' portfolios focused on converting paper documents into equivalent electronic documents. In these early stages of e-portfolio system design, standard setting entailed defining preferred page-formatting protocols, of which the primary choices were Word docs, PDF and html. As the importance of meta-data became clearer, there was a widespread drive to adopt XML. This is the current state of art. Significant benefits, however, appear to reside in a transition away from XML-based e-portfolios to e-portfolio systems based on the Semantic Web and RDF. The reasons are sketched out in a later section.

E-Portfolios: Four Perspectives

E-portfolio systems are shaped by, and affect, multiple constituencies and stakeholders, each with different perspectives. Constituent orientation analysis (Tellefsen, 1995) makes explicit the perspectives of these constituencies and stakeholders. Four key constituencies involved with e-portfolio systems are:

- Educators
- Learners
- E-portfolio system designers
- Education organization managers

From an **educator's** perspective, pedagogic integrity is essential to the design of e-portfolio systems. Educators seek assessment methods that are congruent with the values of curricula as designed, and are fair and efficient to use. E-portfolios enable some automation of assessment management and are especially suitable for constructivist curricula, as discussed in the previous section. Complex portfolios match well with the methods and values of problem-based learning (PBL). Many educators are 'semi-skilled' non-specialist computer users, who are not necessarily aware of the potential benefits and limitations of alternative technical platforms. One of the purposes of this chapter is to raise awareness of pedagogical implications of different technological options.

From the perspective of **learners** e-portfolios offer potential to provide an integrated record of their formal educational achievements within educational institutions, their achievements within formalized continuing professional development programs, their informal learning through on-the-job professional experience, and their learning via self-education. This integrated record of their learning and assessment has value when applying for employment, promotion or further courses of formal study because it provides an organized repository from which examples can be selected according to need.

From an **e-portfolio system designer's** perspective, the breadth of automation possibilities for e-portfolios available via the digital environment offer opportunities to improve institutional educational outcomes, to improve efficiency of assessment, communication, and reporting, and to reduce the time spent on administrative tasks related to educational processes. A criticism of many previous approaches to e-portfolios is that they have neglected pedagogy and learning design and prioritized technology implementation.

Those who own or define the protocols, software systems and structures by which e-portfolios are encoded, managed and accessed have, perhaps inadvertently, taken control of pedagogy. There are also ongoing tensions between commercial interests (e.g. ePortaro, eNVQ, Scioware (Blackboard e-portfolio add-in)) who seek proprietary control of learning platforms, and open source interests (e.g. Open Source Portfolio Initiative (OSPI), ePortConsortium, Learning ePortfolio (http://eportfolio.d.umn.edu/) who define standards within which anyone can develop e-portfolio systems that are interoperable with e-portfolio systems regardless of their proprietary or open-source origins. From the perspective of e-portfolio system design specialists within both camps, architectural issues are central.

The e-portfolio system designer's perspective on e-portfolios focuses primarily on data elements; data types; data relationships and their management, access and presentation, computer systems, computer and networked structures; and systems and communication architectures. In practical terms this includes the choice of data entities and their relationships, protocols, programming languages, data space, database structures, and their implications for hardware issues such as the size of servers, required storage, bandwidth, and server load distribution. Currently, e-portfolio systems design professionals appear to regard e-portfolio elements and architectures in a similar manner to the way learning objects are viewed by the designers of in database-driven e-Learning systems (see, for example, the IMS e-portfolio best practice definition at http://www.imsglobal.org/ep/epv1p0/imsep_bestv1p0.html). In terms of improving pedagogical outcomes for the future, technically minded e-portfolio-systems designers can contribute to the debate on e-portfolios by explaining how new computer technologies may offer improvements to institutional and educational processes.

From the perspective of **education organization management** constituency, e-portfolios help institutions improve the efficiency and effectiveness in fulfillment of institutions' educational purposes, missions and visions. Central roles of education managers involve responsibility for education and teaching provision, and management of associated resources, administration and costs. Managers' aims to maximize economic benefits, improve institutional competitiveness and to reduce cost and strain of resource management, while maintaining educational quality at an acceptable level are achievable. In terms of e-portfolios, the use of structured and automated e-portfolio-based education programs can offer improved course consistency, and improved alignment of learning processes with students' expectations and lifestyle via integrated delivery of learning content and its assessment. E-portfolios can be integrated with other e-learning systems for autodidactic learners.

For the complex of constituencies with an interest in e-portfolios there are some broader positive implications of adopting a more inclusive perspective on e-portfolios that extends their role across institutions. From a learner's perspective, e-portfolios offer improved integration between the multiple different aspects of their learning-focused relationship within and between educational institutions and their life experiences. A more inclusive approach to e-portfolios helps educators, e-portfolio system designers and education organization managers efficiently and effectively integrate the resources necessary to teaching, learning and assessment. From a whole-of-organization perspective, the potential for combining inclusive e-portfolio systems with other institutional systems offers the possibility of improved institutional effectiveness through connections between education and interdependent support processes, and though avoiding inefficient local sub-optimization of different parts of organisations. Later in the chapter it is suggested that W3C Semantic Web technologies such as RDF, OWL, RDF/XML and other components of the defined Semantic Web offer he most appropriate programmatic framework for this integration of e-portfolio systems with other information systems within an education organization

E-Portfolios: Practical Developments

The earliest forms of e-portfolios were limited by the computer skills of educators and computer technology. These replicated paper-based documents using Microsoft Word and similar office software along with simple 'text and image' html pages and the concept of e-portfolios was badly compromised. High status was given to educators claiming to be using electronic portfolios and the temptation of easy access to increased status combined with educators' difficulties of using office software to combine several documents in a single coherent 'electronic container' resulted in a plethora of e-portfolios that were simply single document reports, essays or term papers.

A second wave of e-portfolios used publishing software designed for creation of compound electronic documents. This software included Microsoft 'Binder', the OpenDoc systems of Apple, IBM and WordPerfect Corporation, Envoy from WordPerfect Corporation, and the 'publish and subscribe' model of Apple. In addition, free 'reader' software such as Envoy became available that was embedded in the document and enabled anyone to be able to read a document regardless of whether they had the originating software. Later, Envoy was eclipsed by Adobe's pdf format after Adobe reduced its pricing (the now free Acrobat Reader was originally US$50!) The development of specialized compound document software programs waned as desktop publishing software became more ubiquitous and there was increased awareness that many common office programs such as Microsoft Word, Excel and PowerPoint could incorporate multiple documents and publish them in html on the Internet.

Compound documents began to be published directly on the Web in multiple formats as multiple linked Web pages via user-friendly Web-publishing software such as Macromedia Dreamweaver or Adobe GoLive! Later, this second wave incorporated video and sound files within e-portfolios. Many examples of these early academically and technologically simple approaches to e-portfolios are described by Helen Barrett (see, for example, http://electronic-portfolios.com/ and the table at http://elect-ronicportfolios.com/sys-tems/concerns.html). This style of e-portfolio can be replicated in PowerPoint using slides in collaboration with the integrated notes and the ability to cross-reference, navigate and move within the document using internal URI's (Uniform Resource Indicators).

These simple first and second wave e-portfolios used ordinary office document-creation methods to build e-portfolios as conceptual replicas of paper-based portfolios. The third wave of e-portfolios comprised dynamically composed Web pages, in which documents were pulled from one or more databases held on Web servers, collated in real time and formatted as Web pages. Choices of system characteristics such as information protocol and database type were relatively ad-hoc in this third wave

The fourth and current generation of e-portfolio systems is database driven and characterized by the widespread use of meta-data and XML. A key difference between the third and the fourth generation of e-portfolios is increased use of meta-data about both the data contained within the Web page and the Web page itself. Meta-data is data *about* data. In the early database driven e-portfolios, information about data was typically held in lookup tables within the server-side database. The transition to XML with its focus on meta-data has been used as a means to create a standard platform for e-learning and e-portfolio systems. Typically, recent developments in e-portfolio systems have followed e-learning system models. For example, e-portfolios in Moodle (http://moodle.org/), Blackboard and WebCT are built on and integrated with an over-arching XML database-driven networked learning management system. In e-learning systems the recent focus has centered on learning objects (Alvarado-Boyd, 2003) attached to predefined 'learning object meta-data' (LOM) that can be indexed and queried by a learning object management system (LMS). This e-learning style of database-driven Web-enabled approach is dominant in the e-portfolio literature (see, for example IMS' approach to Learning System Meta-data at http://www.imsglobal.org/metadata/index.html). In these database driven systems, the use of XML enables variables to be associated directly with meta-data as 'tags'. For example, if 'John Smith is the author of a book, then the information that John Smith is an *author* is meta-data about the data 'John Smith' and coded via an <author> tag. Unlike html, XML does not use predefined tags. They have to be defined as part of each application. This provides flexibility but leads to interoperability problems. The key limitation is that XML was designed for managing meta-data of objects in stand-alone commercial online systems in which categories of objects are easily and unambiguously defined as in a sales catalogue. The simple and rigid approach of XML presents significant problems where the definition of the meaning and value of an object (and hence its meta-data) vary dependent on why the object has been chosen and the subjective opinion of the user and where meta-data is shared across heterogeneous system. The most obvious problems appear as poor interoperability.

Interoperability of e-portfolios is compromised where tags are defined inconsistently across applications or documents, and where meta-data is incorrectly allocated to information. The development of successful automated search processes in e-portfolio systems depend on the quality of meta-data tagging of elements of e-portfolios. Each piece of evidence that indicate a student's skills and knowledge may be associated with multiple meta-data tags based on human judgment. This is particularly a problem in XML-based e-learning systems because the classification of learning-related meta-data information is multidimensional and value laden. As a result, the development of large-scale e-learning systems based on XML has required considerable work in establishment of intermediate standards and middleware. At root, these problems derive from weaknesses of system standardization imposed by XML and XML tagging of learning meta-data. An additional problem in using XML for e-portfolios systems is that e-portfolio information can be both 'data-centric' and 'document-centric' making meta-data management difficult. (see, for example http://www.rpbourret.com/xml/XMLAndDatabases.htm).

The fifth, future, generation of e-portfolio system is currently evolving, and we believe is likely to be based on RDF and related protocols from components of the Semantic Web. RDF evolved from XML (http://www.w3.org/XML/1999/XML-in-10-points.html). The primary purpose of RDF is to provide a consistent framework for the generation and management of meta-data. This both reveals and potentially addresses many of the problems associated with XML-based systems that require better management of data using meta-data. This issue is discussed more fully in the next section.

E-Portfolios: Significance of Transition to RDF from XML

The implications for e-portfolio systems of the transition from XML to Semantic Web technologies can be illustrated using tools and concepts from the realm of Systems Analysis. The main differences between XML and RDF concern how they control variety in the system. That is, how and where XML and RDF act to standardize aspects of e-portfolio systems. The standardisation effects of XML occur at the end of the system nearest to the creation of the appearance of Web pages. This is not surprising as its original function was to facilitate large-scale publishing. In contrast, the standardisation actions of RDF occur at the other end of the system where it provides a framework within which systems using meta-data can be developed to maximise their interoperability.

Why is this important? E-Portfolio and e-learning systems are characterized by the need to deal with high levels of variety because learning-related information, knowledge and skill are multidimensional phenomena and human learning and e-portfolio systems contexts are highly variable and complex. In Systems Analysis, a key 'Law' that addresses this phenomenon is Ashby's Law of Requisite Variety (see, for example http://pespmc1.vub.ac.be/REQVAR.html). According to Ashby's Law all this variety has to be controlled somehow and somewhere in each e-portfolio system. E-portfolio management systems must have at least the same amount of variety as the variety of information to be managed. This management variety can, however, be located almost anywhere in the e-portfolio system. Standardisation at lower system levels of data and page description as is done by XML means that 'spare' variety must be accommodated at higher system levels. This typically leads to a proliferation of multiple proprietary and non-proprietary standards in the middle and higher areas of the system in order to control excess variety. In practical terms, spare system variety at higher e-portfolio system levels translates (without a standardizing system such as RDF) to highly heterogeneous variety of different forms of e-portfolio systems, software systems, networks, and other technical, virtual and real institutions systems with incompatibilities between them. The result is interoperability problems.

Consistent standardization at lower levels of e-portfolio systems by XML is problematic and difficult to implement because it requires consistent pre-defined and pre-structured ways of managing meta-data to enable meta-data to be consistently, accurately and unambiguously searchable within e-portfolio systems. In turn, this requires tightly specified, pre-defined and accurately applied meta-data vocabularies defined identically and consistently across all potentially interoperable XML applications and data. This is particularly difficult in a knowledge-based environment such as education where meanings, values, data and meta-data are highly contextualized and dynamic. To achieve interoperability between e-portfolio systems requires standardization at the higher system levels. This is the opposite of what is happening with XML-based approaches to e-portfolios.

From a systems perspective, the solution is straightforward: enable standardization at the higher system levels and allow necessary variety to be accommodated by increased freedom at lower system levels. The systemic weaknesses of XML in controlling variety are substantially resolved by the Semantic Web infrastructure defined by W3C, via the Resource Definition Framework (RDF) and related RDF/XML and OWL, created by the W3C organization for managing semantic meta-data in Web environments. RDF helps controls variety from top down because its role is standardizing how meta-data is managed. It provides a basis for maintaining interoperability of both e-portfolio and e-learning systems because it provides an overarching systems *framework* for managing meta-data and data (RDF = Resource Definition Framework).

RDF provides a consistent simple semantically coherent high-level scalable framework that can be parsed by machines (Bray & Brickley, 2001). It uses *Uniform Resource Identifier references* (URIs) for describing resources in terms of simple properties and property values. A URI can be created to refer to ***anything***:

- Network-accessible things, e.g. electronic documents, network enabled/accessible resources.
- Real things such as humans, businesses, and physical books not accessible via computer networks
- Abstract concepts such as the idea of "book title" and the colour red.

RDF's use of URIs supports e-portfolio system development by allowing e-portfolios to refer to real world or abstract elements that are not network accessible. OWL extends this ability semantically (Ford, 2004; Manola & Miller, 2004) increasing the connection between e-portfolios and real world learning. In addition, RDF allows the physical separation of content of e-portfolios from their metadata by enabling the meta-data used in e-portfolio systems to be 'pointed to' by the use of URIs located *anywhere* on the Internet. It provides integration of different forms of meta-data and graceful resolution where meta-data is inconsistent or missing. Key benefits for the use of RDF in e-portfolio systems are that RDF statements can be parsed by standard machine readers; RDF is scalable; it allows seamless integration of physical, real world physical entities with digital e-portfolios; it supports transferability (interoperability) of e-portfolios between different institutional contexts; and it easily accommodates different qualities and structures of meta-data.

In practical terms anyone can write property and value data about objects, and different vocabularies can be freely mixed in RDF. Instead of providing vocabularies, RDF provides the framework within which individuals and organizations define consistent shared vocabularies. This is another aspect of how the RDF framework reduces variety at higher system levels.

Programmatically, RDF supports a smooth transition to other systems outside the immediate e-portfolio arena because RDF graph-based triples are a machine-processable way of managing entity-relationship data. Hence, RDF can interface with similar information structures such as records in databases (flat file or relational) and rows in spreadsheets. This offers the potential to closely link RDF-based e-portfolio systems with other education administration systems.

E-Portfolios: Integration with Education Administration Systems

The ability of RDF-based e-portfolio systems to refer to real things such as people in non-digital, real world domains provides a bridge, in informatic terms, between e-portfolio systems and other education administration computer systems including human resources systems, research and research support systems, maintenance systems, financial systems, management systems, asset and resource management systems, training and educational support systems, governance systems, marketing & sales systems, customer relationship management systems, evaluation systems, strategy and planning systems. The benefits from linking RDF-based e-portfolio systems with education management systems include:

- Potential for direct synergetic improvements in efficiency and effectiveness. For example, the implement a learning program using e-portfolios also needs integrated connection with human resources management systems, asset management systems, and technology management systems.

- Integration leading to potential for automation, reduced administration, simplified processes, less management overhead, fewer errors and decreased systems failures.
- Potential for system optimization at 'whole of organization' level to facilitate the contribution of e-portfolios to fulfilling vision, mission and strategy agendas of top management through reducing problems caused by local sub- optimization of individual education and learning processes at the expense of broader organizational aims and objectives.

The case for integrating inclusive e-portfolio systems with other systems of education organizations is supported by experience in other fields such as the use of Enterprise Resource Planning systems (ERP) and Product Life Management systems (PLM) which indicate that whole of organization benefits are realized when specialized systems such as e-portfolio systems are integrated into organization-wide systems in the areas of human resources, asset management, room management, accounting, and operational support.

A significant managerial decision for educational organizations planning to use e-portfolio systems (and other e-learning systems) is whether to commit funding to purely XML-based systems with their intrinsic meta-data and interoperability problems or wait until more fully developed RDF-based systems become available. Those developing e-portfolio systems will also have to decide if and when to make the transition to RDF/OWL based Semantic Web technologies as the basis for e-portfolios systems programming and meta-data management.

Future Directions

Obvious future directions are for e-portfolios to emerge as a standard supplement to the curriculum vitae for professionals across many disciplines and the life-long locus of all of a person's biographic, personal, educational and professional information. E-portfolios are becoming a means to improve integration between courses and to recognize prior learning. These uses are likely to expand, facilitating transferability within and between institutions.

Choice of future technology available for e-portfolios may be determined by the relative balance of power between e-portfolio system suppliers. If XML remains the primary means of standardization, then interoperability will occur by adoption of proprietary middleware software protocols in the same way that Microsoft Windows has dominated the desktop market. Currently, dominant players in the e-portfolio arena such as Blackboard are resisted by the Open SourceInitiative Portfolio Initiative (OSPI) (www.osportfolio.org), the best established of the open source portfolio systems at this time. The position of OSPI in e-portfolios is significantly strengthened by its relationship with the Sakai open source e-learning content management system organization (www.sakaiproject.org). Establishment of public interoperability standards such as the Semantic Web by W3C, supported perhaps by the strength of the Open Source movement, may help resolve interoperability issues at a system level. The use of RDF contributes by helping reduce the possibility of commercial monopolistic control of e-portfolios because it reduces the need for proprietary middleware standards.

Summary

There are many new opportunities in education offered by e-portfolios and new forms of e-portfolio systems. Technological platforms to support e-portfolios have become both sophisticated and user-friendly, so it should be possible to give more sustained attention to how technological possibilities of e-portfolios can be best co-opted to pedagogical methods and educational goals.

There are opportunities to extend the use of e-portfolios to improve effectiveness and efficiency learning and assessment processes and to reduce educators' workloads via automation of a variety of administrative processes. There are immediate opportunities for e-portfolios to offer increased benefit to students especially where e-portfolios can improve ease of transfer of evidence for academic achievement between educational institutions and provide students with means to demonstrate non-formal learning.

There is an important opportunity to extend the role for e-portfolio system designers to explain to educators in simple terms the pedagogic opportunities made available by ongoing changes in technologies..

A key challenge for educators is how to best understand and exploit the new possibilities made available by technical change in computer systems and programming protocols. The main technical challenges in the e-portfolio arena are to resolve the interoperability problems; facilitate the transition to Semantic Web technologies that better manage meta-data; and establish the technological foundation for cross-institutional e-portfolio-evidenced education. Some presently unresolved ownership, privacy and ethical issues are likely to emerge where education organizations or related partners provide server space where people store large amounts of personal data that may include data not only about themselves but also about others. For education organization managers and educators there is a difficult choice as to whether to invest in current XML-based e-portfolio systems and later make the transition to RDF-based applications or to move as soon as possible to RDF-based e-portfolios.

In summary, this chapter has addressed several contemporary issues relating to e-portfolios. It has discussed key terms and concepts and provided an overview of e-portfolios and their historical development. The perspectives of four main constituencies were identified as they relate to the-portfolio system design: learners, educators, e-portfolio system designers, and education organization managers. The benefits of integration of e-portfolio systems with other education organization management systems were discussed. Particular attention was paid to interoperability problems and their resolution via Semantic Web technologies such as RDF.

References

Anderson, P., & Blackwood, A. (2004). Mobile and PDA Technologies and their future use in education. *JISC Technology and Standards Watch, 4*(3).

Batson, T. (2002). *The Electronic Portfolio Boom: What's it All About?*, from http://www.campus-technology.com/article.asp?id=6984

Bowden, J., Hart, G., King, B., Trigwell, K., & Watts, O. (n.d). *Generic Capabilities of ATN University Graduates*, 2006, from http://www.clt.uts.edu.au/ATN.grad.cap.project.index.html

Bray, T., & Brickley, D. (2001, 24 Jan 2001). *What is RDF?* Retrieved July 2005, 2005, from http://www.xml.com/pub/a/2001/01/24/rdf.html

Davies, B., & Durkin, M. (1991). 'Skill', 'Competence' and 'Competencies'. *Youth and Policy.*

Ford, P. (2004, May 20, 2004). *Berners-Lee Keeps WWW2004 Focused on Semantic Web.* Retrieved July 2005, 2005, from http://www.xml.com/pub/a/2004/05/20/www-timbl.html

Harman, K., & Koohang, A. (2005). Discussion Board: A Learning Object. *Interdisciplinary Journal of Knowledge and Learning Objects, 1*, 67-77.

Hodgkins, W. (Ed.). (2002). *EEE 1484.12.1-2002 Draft Standard for Learning Object Metadata standard*: IEEE.

Johnson, L. F. (2003). *Elusive Vision: Challenges Impeding the Learning Object Economy (Macromedia White Paper)*, from http://www.nmc.org/pdf/Elusive_Vision.pdf

Koper, R. (2004). Use of the Semantic Web to Solve some Basic Problems in Education: Increase Flexible, Distributed Lifelong Learning, Decrease Teachers' Workload. *Journal of Interactive Media in Education, 2004*(6), 1-9.

Love, D., McKean, G., & Gathercoal, P. (2004). Portfolios to Webfolios and Beyond: Levels of Maturation. *Educause Quarterly, 27*(2).

Love, T., & Cooper, T. (2006). An Inclusive Approach to Learning Object Architectures: Portfolios and RDF. In A. Koohang & K. Harman (Eds.), Learning Objects: Standards, Metadata, Repositories, and LCMS. Santa Rosa: Informing Science Institute.

Love, T., & Cooper, T. (2004). Designing Online Information Systems for Portfolio-Based Assessment: Design Criteria and Heuristics. *Journal of Information Technology Education, 3*, 65-81.

Manola, F., & Miller, E. (Eds.). (2004). *RDF Primer W3C Recommendation 10 February 2004*: W3C.

Mish, F. C. (Ed.). (2004). Merriam-Webster's Collegiate Dictionary (11th ed.). Springfield, MA: Merriam-Webster.

National Learning Infrastructure Initiative. (2003). The Digital Me: Standards, Interoperability, and a Common Vocabulary Spell Progress for E-Portfolios. In *NLII Annual Review* (Vol. 2005). Boulder: National Learning Infrastructure Initiative.

Tellefsen, B. (1995). Constituent Orientation: Theory, Measurements and Empirical Evidence. In B. Tellefesen (Ed.), *Market Orientation* (pp. 111-156). Bergen: Fagbokforlaget.

Treuer, P., & Jensen, J. (2003). Setting Standards for Electronic Portfolios. Vol. 26, No.2, 2003. *Educause Quarterly, 26*(2).

Wiley, D. A. (2001). Connecting learning objects to instructional design theory: A definition, a metaphor, and a taxonomy. In D. A. Wiley (Ed.), *The Instructional Use of Learning Objects* (pp. 1–35). Bloomington: Agency for Instructional Technology.

Authors

Dr Trudi Cooper is a senior lecturer in Youth Work at Edith Cowan University. Last year she gained a prestigious Australian Carrick citation for outstanding contribution to student learning. This was in recognition of her work on the development of portfolio-based assessment methods for practicum assessment across the professions over the past fourteen years. In addition to her curriculum development role within the university, Trudi focuses on the establishment of partnerships with youth and community agencies for collaborative research projects. Previously, Trudi was a youth worker in the UK, where she pioneered feminist youth work with young women, and was involved in development of the national youth work curriculum.

Dr. Terence Love is a research fellow at Curtin University, Western Australia. His undertakes multi-disciplinary research involving social, ethical, environmental and technical factors in design and innovation processes, complex systems, next generation e-learning systems and doctoral education. He is a Fellow of the Design Research Society, a visiting fellow at the Institute of Entrepreneurship and Enterprise Development at the Management School Lancaster University, UK and a visiting professor at UNIDCOM Institute of Design and Communication Research at IADE in Lisbon, Portugal.

Editorial Review Board

Index

A

B

C

D

E

F

G

H

I

K

Q

R

S

Made in the USA
Monee, IL
19 October 2026

66036200R00174